Serving the Reich

THE STRUGGLE FOR THE SOUL OF PHYSICS UNDER HITLER

Serving the Reich

THE STRUGGLE FOR THE SOUL
OF PHYSICS UNDER HITLER

Philip Ball

THE BODLEY HEAD
LONDON

Published by The Bodley Head 2013

2 4 6 8 10 9 7 5 3 1

First published in Great Britain in 2013 by
The Bodley Head
Random House, 20 Vauxhall Bridge Road,
London SW1V 2SA

www.bodleyhead.co.uk
www.vintage-books.co.uk

Addresses for companies within
The Random House Group Limited can be found at:
www.randomhouse.co.uk/offices.htm

The Random House Group Limited Reg. No. 954009

A CIP catalogue record for this book
is available from the British Library

ISBN 9781847922489

The Random House Group Limited supports the Forest Stewardship Council®
(FSC®), the leading international forest-certification organisation.
Our books carrying the FSC label are printed on FSC®-certified paper.
FSC is the only forest-certification scheme supported by the leading
environmental organisations, including Greenpeace.
Our paper procurement policy can be found at
www.randomhouse.co.uk/environment

Typeset in Dante MT by Palimpsest Book Production Limited,
Falkirk, Stirlingshire

Printed and bound in Great Britain by
Clays Ltd, St Ives PLC

Contents

Preface

There is a view becoming increasingly prevalent today that science is no more and no less than a candid exploration of the universe: an effort to find truths free from the ideological dogmas and ambiguities that beset the humanities, using a methodology that is fixed, transparent and egalitarian. Scientists are only human, to be sure, but science (in this view) is above our petty preoccupations – it occupies a nobler plane, and what it reveals is pristine and abstract. This is a time when one can claim without fear of challenge that science is 'disembodied, pure knowledge'. There are scientists and science advocates who consider that historians, philosophers and sociologists, by contrast, can offer little more than compromised, contingent half-truths; that theologians spin webs out of vapour, politicians are venal and penny-pinching vote chasers, and literary theorists are brazen clowns and charlatans. Even the historians, philosophers and sociologists who study science itself are often regarded with suspicion if not outright hostility by practising scientists, not just because they complicate science's tidy self-image but because some scientists cannot imagine why science should need this kind of scrutiny. Why can't scientists be left alone to get on with the business of excavating truth?

This Panglossian description doubtless betrays my scepticism. These trends wax and wane. It is a commonplace to say that scientists once served God, or at other times industry, or national glory. Only a few decades ago science seemed to be happily swimming in the cultural mix, enchanting us with dazzling images of chaos and complexity and looking for dialogue with artists and philosophers. But assaults from religious and political fundamentalists, posturing cultural relativists and medical quacks have understandably left many scientists feeling embattled and desperate to recapture a modicum of intellectual authority. And it remains the case that science has a means of investigation that works and can provide reliable knowledge, and of this its practitioners are fittingly proud.

Yet an insistence on the purity of science is dangerous, and I hope that this book will suggest some reasons for saying so. In studying the responses of scientists working in Germany to the rise of the Third Reich, I could not but be dismayed at how the attitudes of many of them – that science is 'apolitical', 'above politics', a 'higher calling' with a stronger claim on one's duty and loyalties than any affairs of human intercourse – sound close to statements I have heard and read by scientists today.

Peter Debye, who is one of the key figures in this story, was also considered a scientist's scientist. An examination of Debye's life shows how problematic this persona may become when – as is often the case – life calls for something else, something that cannot be answered with a quip or an equation, or worst of all, with the defence that science should pay no heed to such mundane matters.

Debye, like many of his colleagues, doubtless did what he was able in extraordinarily difficult times. Whether or not one feels inclined to criticize his choices, the real problem for scientists in Germany in the 1930s was not a matter of personal shortcomings but the fact the institution of science itself had become an edifice lacking any clear social and moral orientation. It had created its own alibi for acting in the world. We must treasure and defend science, but not at the cost of making it different from other human endeavours, with unique obligations and ethical boundaries – or a unique absence of them.

Debye's story was first brought to my attention by science historian Peter Morris, and he has my deep gratitude for that. My attempts to navigate through the turbulent currents of this particular time and place have been made possible, and hopefully saved from the worst disasters, by the extremely generous help of many experts and other wise voices, and here I am grateful to Heather Douglas, Eric Kurlander, Dieter Hoffmann, Roald Hoffmann, Horst Kant, Gijs van Ginkel, Mark Walker, Stefan Wolff and Ben Widom. Norwig Debye-Saxinger was very gracious in discussing with me some sensitive aspects of his grandfather's life and work. The Rockefeller Archive Center in Tarrytown, New York, made my visit very comfortable and productive.

My agent Clare Alexander, and my editors Jörg Hensgen, Will Sulkin and his successor Stuart Williams at Bodley Head have been as supportive and reliable as I have come, with much gratitude, to anticipate. I am particularly grateful on this occasion for Jörg's perspectives on German culture and history. I was very glad to have benefitted once again from the sensitive and reliable copy-editing of David Milner. As ever, my wife Julia and my family are my inspiration.

Philip Ball
London, March 2013

Introduction:
'Nobel Prize-winner with dirty hands'

Very few great twentieth-century physicists are household names, but Peter Debye must enjoy, if that is the right word, one of the lowest returns of fame within this pantheon. Partly this reflects the nature of his work and discoveries. Albert Einstein, Werner Heisenberg and Stephen Hawking have become regarded, in many respects quite rightly, as pronouncing on deep mysteries about the nature of the physical world. Debye, in contrast, made his largest contributions in an abidingly unfashionable field of science: chemical physics. He decoded the physical character of molecules, and especially how they interact with light and other forms of radiation. His range was remarkable: he helped to understand, for example, how X-rays and electron beams can reveal the shapes and movements of molecules, he developed a theory of salt solutions, he devised a method for measuring the size of polymer molecules. For some of this work he won a Nobel Prize in 1936. He has a scientific unit named after him, and several important equations bear his name. None of this sounds terribly earth-shaking, and in many ways it is not. But Debye is rightly revered by scientists today as someone with phenomenal intuitive insight and mathematical skill, who could see to the heart of a problem and develop its description in ways that were not just profound but useful. It is very rare to find such theoretical and pragmatic sensibilities combined in a scientist.

His colleagues spoke warmly of him; his obituaries were uniformly admiring. He fathered a loving family, and exuded the air of a hale, dependable, outgoing spirit, liking nothing more than a hike or a spell of gardening with his wife. There was, admittedly, nothing unconventional in his character, in the manner of Einstein or Richard Feynman, to snare the imagination – but wasn't that in itself something of a virtue?

So it came as a shock when, in a book called *Einstein in Nederland* published in January 2006 by Dutch journalist Sybe Rispens, Debye was accused of Nazi collusion. In an article written for the Dutch periodical *Vrij Nederland* to coincide with the book's publication, Rispens characterized Debye as a 'Nobel Prize-winner with dirty hands'. He was never a member of the Nazi Party, Rispens admitted, but he was a 'willing helper of the regime' and had contributed to 'Hitler's most important military research program'. Rispens described how, from 1935 until he left Germany at the end of 1939, Debye had been head of the prestigious Kaiser Wilhelm Institute for Physics in Berlin, where subsequently work had been conducted on the military uses of nuclear power. And as the chairman of the German Physics Society in 1938, Debye signed a letter calling for the resignation of all remaining Jewish members of the society – an action that Rispens called 'effective Aryan cleansing'. Even while Debye was in the United States during the war (where he remained at Cornell University in Ithaca, New York, until his death in 1966), he had maintained contact with the Nazi authorities, in Rispens' view keeping open the possibility of returning to his post in Berlin once the hostilities were over.

Debye's conduct in Nazi Germany had previously been presented largely as that of an honest man forced unwillingly into compromises by a vicious regime whose excesses finally drove him into exile. That Debye might have had more selfish motivations was a decidedly unwelcome idea. One commentator argued that this suggestion of hitherto unimagined complexity and controversy in the life of a revered physicist left his admirers feeling 'deprived of a hero'.

It's not clear that Rispens' accusations would have been afforded much attention by scientists, however, had it not been for the response that followed in the Netherlands. Two universities associated with Debye's name panicked and rushed to distance themselves. The Debye Award for Research in the Natural Sciences was instituted in 1977 by Debye's friend, the industrialist Edmund Hustinx, and was administered by the University of Maastricht. In February 2006 the university asked the Hustinx Foundation for permission to drop Debye's name from the award, saying that he 'insufficiently resisted the limitations on academic freedom' during the Nazi era. 'The Executive Board considers this picture difficult to reconcile with the example associated with a naming of a scientific prize', declared a press release from the university. And the University of Utrecht, which hosted the renowned Debye Institute for Nanomaterials Science, likewise announced that 'recent evidence' was 'not compatible with the example of using Debye's name', which would henceforth be dropped from the institute's title.

Those actions contrasted with the response of the chemistry department of Cornell University, which had long been proud to have Debye among its alumni. The department commissioned an investigation into the allegations in collaboration with historian Mark Walker of Union College in Schenectady, a leading authority on German physics during the Third Reich. It concluded that Debye was neither a Nazi sympathizer nor an anti-Semite, and that 'any action that dissociates Debye's name from the [department] is unwarranted'.

Walker and other historians of science insisted that Rispens had given a polarized caricature of Debye which obscured the fact that his response to Nazi rule was no different from that of the vast majority of German scientists. Very few of them actively opposed the Nazis inside Germany – scarcely any non-Jewish professors, for example, resigned their posts or emigrated in protest at Hitler's discriminatory Civil Service Laws of 1933. But by the same token, only a small minority of scientists enthusiastically embraced the poisonous doctrines of the National Socialists. Most scientists in Germany, the historians pointed out, made accommodations and evasions in the face of the intrusions and injustices of the Nazi state: perhaps lodging minor complaints, ignoring this or that directive, or helping dismissed colleagues, while failing to mount any concerted resistance. They were primarily concerned to preserve what they could of their own careers, autonomy and influence. Debye was one of these, no better and no worse than a host of other famous names.

Whatever the merits of Rispens' claims – and I shall examine them in this book – the 'Debye affair' reopened a long-standing and controversial debate about the actions of the German physicists during Hitler's rule. Did they demonstrate any serious opposition to the autocratic and anti-Semitic policies of the National Socialists, or did they on the contrary adapt themselves to the regime? Should we consider these scientists to have occupied a special position, with obligations beyond the quotidian, by virtue of their social and professional roles, their international connections and their scientific and philosophical world views? Was science itself commandeered by the National Socialists for its ideological and military programme? Was it, as some have said, destroyed by the state's racial policies? Or did it survive and in some respects flourish, at least until the bombs began to fall?

One thing is clear: these questions, and the consequent implications for the relationship of science and the state, will not be addressed by the 'persistent and virulent use of the Janus-like combination of hagiography and demonization, the black-and-white characterization of scientists' that Walker feels has often blighted earlier attempts to comprehend science in the Third Reich. There is even now a tendency to present the choices that the scientists in Germany made in straightforward categories of 'right' and 'wrong', which moreover tend to be categories determined by the omniscient hindsight of champions of tolerant liberal democracy. One does not need to be a moral relativist to find dangers in such a position. There are a few heroes and villains in this tale, to be sure. But most of the players are, like most of us, neither of these things. Their flaws, misjudgements, their kindnesses and acts of bravery, are ours: compromised and myopic, perhaps, yet beyond good and evil – and human, all too human.

Three stories

This is true of the three figures examined in this book, whose case histories illuminate, in their contrasts and their parallels, the diverse ways in which the majority of scientists (and other citizens) situated in the grey zone between complicity and resistance adjusted to Nazi rule. It is precisely because Peter Debye, Max Planck and Werner Heisenberg were neither heroes nor villains that their stories are instructive, both about the realities of life in the Third Reich and about the relationship between science and politics more generally. The roles of Planck and Heisenberg have been examined by historians in great detail; Debye has in the past been considered a minor and almost incidental figure, which is precisely why the recent eruption of the

Debye affair is significant. Yet despite the immense amount of research on the German physics community under the Nazis, historians still disagree profoundly and even passionately about how it should be judged.

In the contrasting situations and decisions of Debye, Planck and Heisenberg we can find some context for approaching this question. The lives of the three men intersected and interacted in many ways. Debye and Heisenberg shared the same mentor and worked side by side in Leipzig in the early 1930s. Planck encouraged the careers of both, and they saw him as a father figure and moral beacon. Debye insisted, against the wishes of the Nazis, on naming the physics institute that he headed in Berlin after Planck. When Debye left for the United States after war broke out, Heisenberg was his eventual replacement.

Each of these men was a very different personality. It is clear that none of them was enthusiastic about Hitler's regime, yet all were leaders and guides of German science – managerially, intellectually and inspirationally – and they each played a major part in setting the tone of the physics community's response to the Nazi era. Each of them served the German Reich, both before and during that era, and while that was not the same as serving Hitler, let alone accepting his ideology, none of them seemed able to consider carefully how, or if, there was a distinction. Planck was the conservative traditionalist, a representative of the old Wilhelmite elite who considered themselves to be custodians of German culture. Such men were patriots, confident of their status in society and conscious that their first duty was obedient service to the state. Heisenberg shared Planck's patriotism and sense of civic duty, but lacked his preconceptions about the codes of tradition. For him, the hope for a resurgence of German spirit after the humiliation of the First World War lay with a youth movement that celebrated a romantic attachment to nature, to comradeship and frank engagement with philosophical questions. Just as Heisenberg had no qualms about shaping the revolutionary quantum theory, which Planck had reluctantly helped to launch, into a world view that cast doubt on all that went before, so he felt little allegiance to the conservatism of Prussian culture. And Debye is the outsider, who carved out an illustrious career in Germany while steadfastly refusing German citizenship. Faced with the interference and demands of the National Socialists, Planck fretted and prevaricated. Heisenberg sought official approval while refusing to recognize the consequences of his accommodations. Debye is in many ways the most ambiguous of the trio, not because he was the most cunning but perhaps because he was a simpler, less reflective man: the 'scientist's scientist', truly 'apolitical', for better or worse, in his devotion to his research.

The cases of these three men have much to tell us about the factors behind the dominance of the Nazi state. Such a regime becomes possible not because people are powerless to prevent it, but because they fail to take effective action – indeed, even to perceive the necessity of doing so – until it is too late. It is for this reason that judging Planck, Heisenberg and Debye should not be concerned with whether a person's historical record can be deemed 'clean' enough to honour them with medals, street names and graven images. It is about whether we can adequately understand our own moral strengths and vulnerabilities. As Hans Bernd Gisevius, a civil servant under Hitler and a member of the German Resistance, puts it:

> One of the vital lessons that we must learn from the German disaster is the ease with which a people can be sucked down into the morass of inaction; let them as individuals fall prey to overcleverness, opportunism, or cowardliness and they are irrevocably lost.

1 'As conservatively as possible'

Science was done differently a hundred years ago. To appreciate just how differently, you need only compare the traditional group photographs of today's scientific meetings with that from the 1927 Solvay conference on quantum physics in Brussels.* There are no casual clothes here, no students, and most definitely no cheerful grins – only Heisenberg's nervous, boyish smile comes close. The rigidity of the dress code matches the severity of the gazes, which exude an oppressive expectation that codes of conduct will be observed and hierarchy respected. One feels that Hendrik Lorentz, on Einstein's right in the front row, is silently reprimanding us for some breach of protocol. It is, needless to say, an all-male assembly, except for Marie Curie, not yet quite sixty but already looking aged by exposure to the radioactivity that would kill her seven years later. There on the far left of the middle row, stiff and uncomfortable, is Peter Debye.

Much of this appearance simply reflects the times, of course. But some is specifically German, for German-speakers dominate this assembly. Even now German science retains something of this sense of decorum and form; foreign visitors are surprised to find that even close colleagues address one another by title and surname, while grades of seniority are demarcated almost as subtly as they are in Japanese society. And of course the status of personal relationships remains explicitly codified in the *du/Sie* distinction. For the German-speaking scientists at the Solvay meeting this linguistic etiquette reflected one's professional standing – despite being friends by any other standard, the

---

* These invitation-only meetings, usually taking place every three years at the grand Hotel Metropole in Brussels, were sponsored by the Belgian industrialist Ernest Solvay, who had been persuaded to lend his support in 1910 by physicists Walther Nernst and Hendrik Lorentz.

The delegates at the 1927 Solvay conference in Brussels, officially titled 'Electrons and photons'. From left to right: top row, A. Piccard, E. Henriot, P. Ehrenfest, E. Herzen, Th. de Donder, E. Schrödinger, J. E. Verschaffelt, W. Pauli, W. Heisenberg, R. H. Fowler, L. Brillouin; middle row, P. Debye, M. Knudsen, W. L. Bragg, H. A. Kramers, P. A. M. Dirac, A. H. Compton, L. de Broglie, M. Born, N. Bohr; front row, I. Langmuir, M. Planck, M. Curie, H. A. Lorentz, A. Einstein, P. Langevin, Ch.-E. Guye, C. T. R. Wilson, O. W. Richardson.

young Heisenberg and Wolfgang Pauli were *Sie* to one another until they both became full professors.

It is not just unfair but in fact meaningless to evaluate the German physicists' response to Hitler without taking into account the social and cultural expectations that framed it. What today's sneakers and sweatshirts are perhaps telling us is that, among other things, academic scientists no longer enjoy quite the same status as they did when Einstein and his peers lined up soberly for posterity's sake at the Hotel Metropole.

That respect brought with it duties and responsibilities. German academics came largely from the middle and upper middle classes: they knew their niche in the social hierarchy and that, by occupying it, they were obliged to support the tiers. The education that these men received placed great emphasis on the concept of *Bildung*, a notion of development that went far beyond the matter of learning facts and skills. It entailed cultivation and maturation of personality – intellectual, social and spiritual – in the course of which the individual learnt to align his outlook with the demands and expectations of society. The German education system stressed the importance of philosophy and literature, bestowing an appreciation for *Kultur*; the educated elite were expected to be guardians of this national heritage, a role for which they felt in a sense contracted by the state. The Dutch physicist Samuel Goudsmit, who as we shall see had good reason to ponder on the consequences of German scientific culture in the early twentieth century, wrote in 1947 that 'Prussia . . . could not afford more than a qualified liberty for its own bourgeoisie, and could certainly not afford to breed men of science who might question the divine mission of the State.'

This form of patriotic devotion was not, however, seen as a political stance, but as something that superseded it. 'Like the majority of the professoriate', says historian Alan Beyerchen, 'German physicists desired strongly to remain aloof from political concerns.' This does not mean that they spurned politics altogether. Most respectable citizens proclaimed an allegiance to a political party – but they did so *as* citizens, and generally maintained a clear separation between the political and the professional. It was precisely the complaint often made against Einstein, and even conceded by some of his supporters, that he did not respect this division – that he 'played politics' through his advocacy of internationalism. His pacifism, which was part and parcel of that attitude, made him still more suspect, for patriotism and national pride were regarded not as a choice but as a duty. In striking contrast to what one might anticipate from academics today, there was scarcely any support from scientists for the popular left-wing Bolshevik movements in the aftermath of the First World War. On the contrary, the German university faculties were predominantly of a conservative inclination, opposed to the Weimar government and resentful about the war reparations.

Physics, a young discipline less steeped in tradition than most others, was somewhat more liberal – but again we must not assume that this has quite the same connotation as today. The allegedly apolitical stance of German academics was in fact tailored to suit a particular political position: it was 'apolitical' to observe the convention of supporting German militarism and patriotism, and equally so to be antagonistic towards democratic Weimar.

The reluctant revolutionary

No one illustrates the traditionalist traits of the *fin de siècle* German scientist better than Max Planck. According to his biographer John Heilbron, 'Respect for law, trust in established institutions, observance of duty, and absolute honesty – indeed sometimes an excess of scruples – were hallmarks of Planck's character.' These were his great strengths; they are the reasons why we must consider him an honourable man. In the Nazi era they would also become weaknesses, trapping him into stasis and compromise.

Born in 1858 in Kiel, Holstein, when it was still officially Danish, Planck was a gentle man; as he put it himself, 'by nature peaceful and disinclined to questionable adventures'. The finest adventure that he could conceive of was one removed from the messy, unpredictable travails of human community: science. 'The outside world is something independent from man,' Planck wrote, 'something absolute, and the quest for the laws which apply to this absolute appeared to me as the most sublime scientific pursuit in life.' Like many scientists today, Planck seemed to find and welcome in science an abstract order that made few demands on the human soul. His relationships did not lack warmth, to judge by the affection that he inspired, but they were conducted with great reserve and decorum: only with people of his own rank could Planck relax a little and enjoy a cigar.

But this mild nature did not prevent a certain bellicosity when it came to national pride and sentiment. Accepting the standard view that Germany was engaged in a purely defensive struggle at the outbreak of the First World War, he wrote to his sister in September 1914 to say 'What a glorious time we are living in. It is a great feeling to be able to call oneself a German.'

Taken in isolation, such a comment might be seen as evidence that Planck was a nationalistic chauvinist. And if one can make that charge of Planck, who his colleagues praised in 1929 for 'the spotless purity of his conscience', there would then be hardly a German scientist of that age who could not be similarly labelled. Indeed, one could

Max Planck (1858–1947) in 1936.

strengthen the charge in several ways. Planck was one of the many scientists who signed the infamous Professors' Manifesto, 'Appeal to the Cultured People of the World', in October 1914, supporting the German military action and denying the (all too real) German atrocities perpetrated in occupied Belgium. Here Planck joined his name to those of the chemists Fritz Haber, Emil Fischer and Wilhelm Ostwald and the physicists Wilhelm Wien, Philipp Lenard, Walther Nernst and Wilhelm Röntgen, existing or future Nobel laureates all (but not, notably, Einstein). More, Planck supported the moderate right German People's Party (Deutsche Volkspartei, DVP), in which it was not hard to find currents of anti-Semitism. He was sceptical of the political validity of democracy in the modern sense.

But it would be dishonest to select Planck's character for him in such a manner, for we might equally highlight his progressive, enlightened attitudes. He supported women's rights to higher education (although not universal suffrage). He refused to sign an appeal drawn up by Wilhelm Wien in 1915 which deplored the influence of British physicists in Germany, accused them of all manner of professional transgressions, and called for scientific relationships with England to be severed. And Planck had the courage to realize his error in putting his name to the Professors' Manifesto and to recant publicly during the war. It is some kind of testimony that Einstein came to hold Planck in close affection and esteem, and that the part-Jewish physicist Max Born said of him that 'You can certainly be of a different opinion from Planck's, but you can only doubt his upright, honourable character if you have none yourself.' We need to know all this before we see what became of Planck, and then of his name.

Planck's characteristics were reflected in his science, which was cautious, conservative and traditional yet displayed open-mindedness and generosity. He readily admitted that he was no genius – indeed, it has been said that he was so often wrong, it was not surprising he was sometimes right. But he made one great discovery, and it brought him a Nobel Prize in 1918.* It concerned a question that seems simultaneously exceedingly esoteric and mundane: how radiation is emitted from warm bodies. What it led to was quantum theory.

So-called 'black-body radiation' – the electromagnetic radiation (including light) emitted by a warm, perfectly non-reflective object – was a long-standing puzzle. Atomic vibrations in the object make its electrons oscillate – and as the Scottish physicist James Clerk Maxwell had shown in the mid-nineteenth century, an oscillating electrical charge radiates electromagnetic waves. The hotter the atoms, the faster they vibrate, and the higher the frequency (shorter the wavelength) of the emitted radiation.†

Towards the end of the nineteenth century, Wien had found by experiment the mathematical relationships between the temperature of a 'black body', the amount of energy it radiates, and the wavelength of the most intense radiation. This wavelength gets shorter as the temperature increases, an observation familiar from experience with an

---

* It is sometimes said that Planck made *two* great discoveries – the second being Einstein.

† Light consists of simultaneously vibrating electrical and magnetic fields: it is electromagnetic radiation. Visible light has wavelengths ranging from around 700 millionths of a millimetre (red) to around 400 millionths (violet). The longer the wavelength, the lower the frequency of the vibrations.

electric heater: as it warms up, it first emits long-wavelength, invisible infrared rays (which you can feel as heat), then red light and then yellow. Objects hotter still acquire a bluish glow. In attempting to explain this process of emission from the warm, vibrating atoms of the black body, Planck stumbled on the quantum nature of the physical world.

Previous efforts to relate atomic vibrations to temperature seemed to lead to the conclusion that the amount of energy radiated should get ever greater the shorter the wavelength of the radiation. In the ultraviolet range (that is, at wavelengths shorter than that of violet light) this quantity was predicted to rise towards infinity, an evident absurdity called the ultraviolet catastrophe. In 1900 Planck found that the equations of black-body radiation would produce more sensible results if one assumed that the energy of the 'oscillators' in the black body were divided into packets or 'quanta' containing an amount of energy proportional to their frequency. He labelled the constant of proportionality h, which became known as Planck's constant.

For Planck this was simply a mathematical trick – as he put it, a 'fortunate guess' – to make the equations yield a meaningful answer. But Einstein saw it differently. In 1905 he argued not only that one might assume Planck's energy quanta to be real, but that they applied to light itself: he wrote that the energy in light 'consists of a finite number of energy quanta localized at points of space that move without dividing, and can be absorbed or generated only as complete units'. These light quanta became known as photons.

Einstein explained that his proposal might be tested by investigating the photoelectric effect, in which light shining on a metal can eject electrons and thereby elicit a tiny electric current. Philipp Lenard had studied the effect closely, and had puzzled over why, as the light becomes more intense, the electrons don't get kicked out of the metal with increasing energy, as one might have expected. But in Einstein's picture, in which the light is composed of photons whose energy is governed by Planck's law, making the light more intense doesn't alter the photons' individual energy; it merely supplies them in greater numbers. This, in turn, increases the number of ejected electrons but not their energies. Only by using light of a shorter wavelength, meaning that the photons have more energy, could the energy of the ejected electrons be increased. Einstein's theory led to predictions that were experimentally confirmed a decade later by the American Robert

Millikan. This work on the photoelectric effect was cited as the primary motivation for awarding Einstein the Nobel Prize in Physics in 1921.

It is hard to overestimate the disruption that Einstein's 'quantum light' paper caused. No one had previously questioned the view that light was a smooth wave, and it is often forgotten now how challenging the notion of 'granular light' was. Even after most physicists were willing to accept a quantum picture of the energies of atoms and their constituent particles, invoking it for light was deemed a step too far, and – despite Millikan's work – it was resisted for two decades.

Planck himself was initially too disturbed by this dislocation in the traditional view of light to accept the quantum hypothesis that he'd unwittingly unleashed. He advised that his constant h, the finite measure of how fine-grained the world was, be introduced into theory 'as conservatively as possible'. Planck came only gradually and reluctantly to recognize that the quantum hypothesis was the best way to understand the world of 'electrons and photons' that he and his peers debated in Brussels in 1927. And yet his broader question – how much of quantum theory is a mathematical formalism and how much reflects physical reality – remained contentious, and is no less so today.

Planck was more receptive to Einstein's second revelation in 1905: the theory of special relativity. Here Einstein proposed that time and space are not uniform everywhere but can be distorted by relative motion. For an object moving relative to another at rest, space is compressed in the direction of motion while time slows down. This mutable notion of what became known as space–time compromised the old view of mechanics based on Isaac Newton's laws of motion, in which the physical world was regarded as a system of bodies interacting with one another on a fixed, eternal grid of time and space. Einstein's discovery was extremely disorientating; literally, it deprived physics of its bearings. Out of the theory of special relativity came a succession of revolutionary concepts: that no object can travel faster than light, that an object's mass increases as it speeds up, that energy and mass are related via the iconic equation $E = mc^2$.* The startling

* The genealogy of this equation is complex – an equivalence of energy and mass was long suspected, and in fact it does *not* rely explicitly and uniquely on the theory of relativity – see http://physicsworld.com/cws/article/news/2011/aug/23/did-einstein-discover-e-equals-mc-squared.

consequences of special relativity are barely apparent, however, until the velocities of objects approach the speed of light – about 300,000 kilometres per second. Scientists could not yet knowingly induce such awesome speeds artificially. That was soon to change.

Planck was an enthusiastic advocate of special relativity, but he was much more wary of Einstein's extension of these ideas in 1912 in the theory of *general* relativity. By apparently dispensing with the force of gravity, reducing it to a distortion of space–time itself, Einstein seemed to Planck to be departing too far from convention and entering into pure speculation. This initial resistance by one of the most eminent German scientists of the age was a source of immense frustration for Einstein. There was rather less hesitancy outside Germany, albeit perhaps for complex reasons. The English astronomer Arthur Eddington was almost zealous in his determination to validate the theory: a pacifist Quaker, he saw this as a way of welcoming German science back into the international fold after the rupture of the war. Eddington has been accused of being selective with the data taken during two expeditions in 1919 to Brazil and Africa to view the solar eclipse and search for bending of starlight round the sun, which general relativity predicted. Whether they were secure or not, Eddington's findings, published the following year, were taken as confirmation of Einstein's genius, and they made him an international celebrity.

Rebuilding German science

The names of scientists working in the German-speaking nations in the early twentieth century – Planck, Einstein, Heisenberg, Schrödinger – are so intimately tied to the revolutions taking place in theoretical physics that it is easy to overlook how precarious German science was at the time. The First World War brought not only a crippling financial burden which eventually ballooned into the hyperinflation and economic stagnation of the Weimar Republic, but also a sense of national shame and isolation. Everyone in Germany felt this affliction; the scientists, accustomed to pre-war supremacy, experienced it especially keenly. The theoretical discoveries of Planck and Einstein at the start of the century had followed on the heels of an unmatched mastery of experimental physics. In 1895 Wilhelm Röntgen at Würzburg had amazed and enchanted the world by discovering X-rays, a form of electromagnetic radiation with very short wavelengths. His work built on the pioneering studies of Philipp Lenard at Heidelberg on 'cathode rays', which were revealed in 1897 to be not rays at all but streams of subatomic, electrically charged particles subsequently called electrons, fundamental constituents of atoms.

The scientific pre-eminence of Germany before the war was not limited to physics. It had dominated chemistry throughout the nineteenth century, thanks to pioneers such as Justus von Liebig, Friedrich August Kekulé, Adolf von Baeyer and August Wilhelm von Hofmann. The German chemists displayed an enviable aptitude for converting their laboratory discoveries into the mass products of a thriving chemical industry. Dyestuffs, pharmaceuticals, fertilizers and photographic products were the mainstay of powerful German industrial companies such as Hoechst, Bayer, BASF and Agfa. At the start of the twentieth century, Emil Fischer at the University of Berlin (Nobel laureate 1902) was arguably the world's foremost organic chemist, while physical chemistry was dominated by Wilhelm Ostwald at Leipzig (Nobel laureate 1909). In physiology, Wilhelm Roux, Hans Spemann and Hans Driesch had made embryology a true science, and the controversial zoologist Ernst Haeckel at Jena had spread the word of Darwinism throughout Germany. In medicine, Robert Koch at Berlin pioneered the understanding of tuberculosis; his one-time assistant Paul Ehrlich helped to launch synthetic pharmaceuticals with the anti-syphilis drug Salvarsan.

Yet even before the First World War, concerns were expressed that German science was in danger of losing its dominant position. In 1909 a seemingly unlikely champion of science, the theologian and historian Adolf von Harnack, argued that the appearance of privately funded scientific institutions in the United States, such as the Carnegie Institution in Washington DC, might leave Germany in the shade. That kind of private enterprise might work for the Americans, but it was not in the German tradition. At the end of the nineteenth century the minister for university affairs, Friedrich Althoff, proposed that a state-funded 'colony' of scientific institutes be set up in Berlin-Dahlem – a kind of German Oxford, affiliated with the universities but independent of them.

This plan cohered in 1911 with the formation of the Kaiser Wilhelm Society (Kaiser-Wilhelm-Gesellschaft, KWG), of which

Harnack was the first president. It was funded partly by industry and partly by the government, and was intended to foster both pure and applied scientific research in an environment that freed the scientists from teaching responsibilities. In contrast to the universities, appointments to the KWG institutes were determined not by the state but by the scientists themselves – the state ministries simply rubber-stamped the decisions. This separation from the university system was to prove critical to the KWG's operation during the Nazi era, for it meant that, unlike university professors, staff at the society's institutes were in general not formally state-employed civil servants. The KWG evolved into a semi-private* organization with over thirty institutes by the end of the 1920s.

The first two of these research centres, the Kaiser Wilhelm Institutes for Chemistry (KWIC) and Physical Chemistry (KWIPC), were opened in Berlin-Dahlem in 1912 by the emperor in person. The institutes that followed were primarily biological and medical: botany, zoology, microbiology, physiology. Physics was not a priority. The precedence awarded to chemistry reflected its industrial importance; the KWIPC was financed by the Jewish banker and entrepreneur Leopold Koppel, a senator of the KWG. Koppel made this endowment contingent on the institute's director being the Jewish German chemist Fritz Haber, who had demonstrated the importance of chemistry for industry and agriculture by developing, between the mid-1890s and 1913, a catalytic process for turning atmospheric nitrogen into fertilizer. Haber's method, which won him a Nobel Prize in 1918, was modified for industrial-scale production by Carl Bosch at BASF. Bosch went on to win a Nobel in 1931 for his work on chemical processing at high pressures, and he became president of the KWG in 1937. The First World War lent fresh significance to the Haber–Bosch process, which was given over largely to the production of nitrogen-rich explosives rather than fertilizer. It has been said that, without this chemical technology, the war would have been over in a year through lack of munitions. At the KWIPC Haber undertook wartime research on the production of chlorine and other poisonous gases for chemical warfare. By 1917

* Harnack claimed somewhat perplexingly that the institutes were 'private institutes and state institutes at the same time'. In 1928 the KWG became officially administered by an autonomous scientific council.

the institute (which is today named after Haber) had grown to house 1,500 personnel, including 150 scientists.

But the war and the political instabilities of its aftermath severely disrupted the aspirations of the KWG and threatened to choke this attempt to revitalize German research. 'At the moment the outlook for our German science is very bleak,' Planck wrote in 1919. 'But I cling strongly to the hope that it will again reach the top . . . if only we can get through the next difficult years decently.' Planck could have no inkling just how difficult those years would become, nor how hard it would be to retain decency.

After the humiliation of the war, supporting science was not just a desirable economic investment but also a way of regaining national prestige. And how deeply humiliating it was. The harshly punitive Treaty of Versailles compelled Germany to pay the fantastic sum of 269 billion gold marks (later reduced, although still it fuelled German hyperinflation), stripped it of territories in Alsace, Upper Silesia, North Schleswig and elsewhere, deprived it of its colonies, allowed it only a tiny army and almost no navy, and excluded it from the League of Nations. The treaty also undermined the support within Germany for the liberal Weimar government that had brokered it.

Unlike Einstein, most scientists and academics responded to this disgrace by turning inward, attempting to salvage some pride by asserting the moral superiority of German culture. The nation might have been broken and humbled by the war, Planck told the Prussian Academy of Sciences in 1918, 'but there is one thing which no foreign or domestic enemy has yet taken from us: that is the position which German science occupies in the world'.

This nationalism, often bordering on chauvinism, was in part a defensive reaction to a vindictive international boycott on German science and scientists after the war. The newly formed International Research Council decided to exclude Germans and Austrians from its committees, meetings and projects. In Britain and the United States the wisdom of this counterproductive gesture was questioned in the 1920s, and by 1926 the council was persuaded to open its doors again to the Germans. Stung by the preceding snub, they refused the invitation. Instead of tempering its cultural isolation by seeking to engage in international affairs, Germany became yet more nationalist and isolationist, insisting stridently on the uniqueness and primacy of

German science. Science, Planck insisted, 'just like art and religion, can in the first instance grow properly only on national soil. Only when such a basis has been established is a fruitful union of the nations in high-minded competition possible.'

Stripped of political power, German leaders and researchers sought to substitute scientific prestige in its place. Even before the war, Harnack's report calling for the establishment of specialized research institutes had been viewed as a quasi-military political strategy, being summarized thus by the Prussian Ministry of Education and the Reich Interior Ministry:

> For Germany the maintenance of its scientific hegemony is just as much a necessity for the state as is the superiority of its army. A decline in Germany's scientific prestige reacts upon Germany's national repute and national influence in all other fields, leaving entirely out of the account the eminent importance for our economy of a superiority in particular fields of science.

This being so, it was the duty of German scientists to act as ambassadors for their country: to impress on the world the strengths and virtues of German science. Einstein's internationalism, which claimed that science was an enterprise without borders and independent of one's country or creed, was considered unpatriotic and distasteful.

When Planck pronounced his gloomy prognosis in 1919, the fissiparous Weimar government was sailing towards economic disaster. Within just a few years, hyperinflation had made nonsense of the mark and the country stood on the brink of total dissolution. In 1923 the cost of a loaf of bread rose into the millions of marks; what could be purchased when you got your wages could become unaffordable by the time you got to the shops. Planck once found, while on a trip as secretary of the Prussian Academy of Sciences, that the money he had been given for expenses when he set off on the train was not enough by the time he arrived at his destination to cover the cost of a hotel room for the night, forcing the 65-year-old to sit up all night in the station waiting room.

In those circumstances, where could money be found to keep German science alive, let alone to restore it to pole position? The

KWG* was compelled to go begging to the Prussian state. In 1920 Harnack and Planck, who had been elected to the society's senate in 1916, enlisted the support of the former culture minister of Prussia to establish the Emergency Association of German Science, an organization that would gather funds for research. Although some money was granted by the state and some by industry, a fecund source was identified abroad. The Rockefeller Foundation in the United States, founded in 1913 by the industrialist and philanthropist John D. Rockefeller, rose above the international boycott of German science to honour its declared intent of promoting 'the well-being of mankind throughout the world', and it entered into negotiations to realize Harnack's vision of creating a nucleus of scientific institutes.

Wind of change

The Weimar government was never less popular than in the early 1920s, leading even liberals to express some nostalgia for the more authoritarian culture of imperial rule. Bavaria had been particularly fragile politically since the end of the war, wavering between extremes. The far-left Independent Socialists led by Kurt Eisner had gained control in 1918, but were inept at governance, and elections the following year handed a majority to the right-wing Bavarian People's Party. When Eisner was shot by a far-right extremist in February 1919, there was fighting on the streets of Munich. An unusually cold winter, in which snow persisted until May, exacerbated the shortages of food and fuel. The unrest continued until 1923, culminating with an attempted putsch against the local government by the National Socialist German Workers' Party (Nationalsozialistische Deutsche Arbeiterpartei, NSDAP), led by Adolf Hitler. The uprising was suppressed and its ringleader imprisoned, but not before Munich was shaken by more violence. During his prison sentence, Hitler spelt out his vision of political struggle:

---

* The continued use of the imperial name in a republic might seem incongruous. But a proposal by some leftist elements that it be changed after the war was strongly resisted – an indication of the conservatism and adherence to tradition of most of its members. Only in 1948 was the name finally changed; we will see later how that came about.

The nationalization of the great masses can never take place by way of half measures, by a weak emphasis upon a so-called objective viewpoint, but by a ruthless and fanatically one-sided orientation as to the goal to be aimed at . . . One can only succeed in winning the soul of a people if . . . one also destroys at the same time the supporter of the contrary.

After the unrest in Bavaria (and elsewhere) dissipated, the Weimar government was granted a brief respite from its travails. The economy at last began to settle, and the worst fears of the middle and upper classes – that there would be a Communist revolution – failed to materialize. This was the 'golden age' that the Weimar era rather selectively evokes in the popular image today: the time of the Bauhaus, jazz, artistic and sexual permissiveness. That was perhaps how it seemed to Berlin bohemians, but very few academics and scientists partook of this hedonistic culture, which they tended to regard with the suspicion and contempt of the elite for the vulgar.

The period of grace ended in 1930, when the federal elections exposed the schism between the creeping extremes of German political life. The National Socialists enjoyed a surge in support, gaining 18 per cent of the vote: 107 of the 577 seats in the German parliament (Reichstag), compared to just twelve in the elections two years earlier. The Social Democratic Party retained its majority, but with only thirty-six more seats than the Nazis, and was hindered by disagreements with the third-placed Communist Party. Thus the Social Democrats could claim no mandate; they could barely govern at all. In the political chaos that followed, support for the National Socialists blossomed. Increasingly they seemed the only party capable of exercising firm rule. Hitler blamed the turmoil on the Jewish bankers and Communist agitators. Naked anti-Semitic sentiment rose like dross to the surface.

That year Adolf von Harnack died, and Max Planck was elected president of the KWG. He thereby became the de facto figurehead of German science, its captain against the gathering storm.

2 'Physics must be rebuilt'

Quantum theory, with its paradoxes and uncertainties, its mysteries and challenges to intuition, is something of a refuge for scoundrels and charlatans, as well as a fount of more serious but nonetheless fantastic speculation. Could it explain consciousness? Does it undermine causality? Everything from homeopathy to mind control and manifestations of the paranormal has been laid at its seemingly tolerant door.

Mostly that represents a blend of wishful thinking, misconception and pseudoscience. Because quantum theory defies common sense and 'rational' expectation, it can easily be hijacked to justify almost any wild idea. The extracurricular uses to which quantum theory has been put tend inevitably to reflect the preoccupations of the times: in the 1970s parallels were drawn with Zen Buddhism, today alternative medicine and theories of mind are in vogue.

Nevertheless, the fact remains that fundamental aspects of quantum physics are still not fully understood, and it has genuinely profound philosophical implications. Many of these aspects were evident to the early pioneers of the field – indeed, in the transformation of scientific thought that quantum theory compelled, they were impossible to ignore. Yet while several of the theory's persistent conundrums were identified in its early days, one can't say that the physicists greatly distinguished themselves in their response. This is hardly surprising: neither scientists nor philosophers in the early twentieth century had any preparation for thinking in the way quantum physics demands, and if the physicists tended to retreat into vagueness, near-tautology and mysticism, the philosophers and other intellectuals often just misunderstood the science.

This penchant for pondering the deeper meanings of quantum

theory was particularly evident in Germany, proud of its long tradition of philosophical enquiry into nature and reality. The British, American and Italian physicists, in contrast, tended to conform to their stereotypical national pragmatism in dealing with quantum matters. But even if they were rather more content to apply the mathematics and not wonder too hard about the ontology, these other scientists relied strongly on the Germanic nations for those theoretical formulations in the first place. Germany, more than any other country, showed how to turn the microscopic fragmentation of nature into a useful, predictive, quantitative and explanatory science. If you were a theoretical physicist in Germany, it was hard to resist the gravitational pull of quantum theory: where Planck and Einstein led, Arnold Sommerfeld, Peter Debye, Werner Heisenberg, Max Born, Erwin Schrödinger, Wolfgang Pauli and others followed.

This being so, it was inevitable that the philosophical aspects of quantum physics should have been coloured by the political and social preoccupations of Germany. As we shall see, it was not the only part of physics to become politicized. These tendencies rocked the ivory tower: the kind of science you pursued became a statement about the sort of person you were, and the sympathies you harboured.

Unpeeling the atom

The realization that light and energy were granular had profound implications for the emerging understanding of how atoms are constituted. In 1907 New Zealander Ernest Rutherford, working at Manchester University in England, found that most of the mass of an atom is concentrated in a small, dense nucleus with a positive electrical charge. He concluded that this kernel was surrounded by a cloud of electrons, the particles found in 1897 to be the constituents of cathode rays by J. J. Thomson at Cambridge. Electrons possess a negative electrical charge that collectively balances the positive charge of the nucleus. In 1911 Rutherford proposed that the atom is like a solar system in miniature, a nuclear sun orbited by planetary electrons, held there not by gravity but by electrical attraction.

But there was a problem with that picture. According to classical physics, the orbiting electrons should radiate energy as electromagnetic rays, and so would gradually relinquish their orbits and spiral into the nucleus: the atom should rapidly decay. In 1913 the 28-year-old Danish physicist Niels Bohr showed that the notion of quantization – discreteness of energy – could solve this problem of atomic stability, and at the same time account for the way atoms absorb and emit radiation. The quantum hypothesis gave Bohr permission to prohibit instability by fiat: if the electron energies can only take discrete, quantized values, he said, then this gradual leakage of energy is prevented: the particles remain orbiting indefinitely. Electrons *can* lose energy, but only by making a hop ('quantum jump') to an orbit of lower energy, shedding the difference in the form of a photon of a specific wavelength. By the same token, an electron can gain energy and jump to a higher orbit by absorbing a photon of the right wavelength. Bohr went on to postulate that each orbit can accommodate only a fixed number of electrons, so that downward jumps are impossible unless a vacancy arises.

It was well established experimentally that atoms do absorb and emit radiation at particular, well-defined wavelengths. Light passing through a gas has 'missing wavelengths' – a series of dark, narrow bands in the spectrum. The emission spectrum of the same vapour is made up of corresponding bright bands, accounting for example for the characteristic red glow of neon and the yellow glare of sodium vapour when they are stimulated by an electrical discharge. These photons absorbed or emitted, said Bohr, have energies precisely equal to the energy difference between two electron orbits.

By assuming that the orbits are each characterized by an integer 'quantum number' related to their energy, Bohr could rationalize the wavelengths of the emission lines of hydrogen. This idea was developed by Arnold Sommerfeld, professor of theoretical physics at the University of Munich. He and his student Peter Debye worked out why the spectral emission lines are split by a magnetic field – an effect discovered by the Dutch physicist Pieter Zeeman in work that won him the 1902 Nobel Prize.*

But this was still a rather ad hoc picture, justified only because it seemed to work. What are the rules that govern the energy levels of electrons in atoms, and the jumps between them? In the early 1920s

---

* This Zeeman effect is the magnetic equivalent of the line-splitting by an electric field discovered by the German physicist Johannes Stark – see page 88.

Max Born at the University of Göttingen set out to address those questions, assisted by his brilliant students Wolfgang Pauli, Pascual Jordan and Werner Heisenberg.

Heisenberg, another of Sommerfeld's protégés, arrived from Munich in October 1922 to become Born's private assistant, looking as Born put it 'like a simple farm boy, with short fair hair, clear bright eyes, and a charming expression'. He and Born sought to apply Bohr's empirical description of atoms in terms of quantum numbers to the case of helium, the second element in the periodic table after hydrogen. Given Bohr's prescription for how quantum numbers dictate electron energies, one could in principle work out what the energies of the various electron orbits are, assuming that the electrons are held in place by their electrostatic attraction to the nucleus. But that works only for hydrogen, which has a single electron. With more than one electron in the frame, the mathematical elegance is destroyed by the repulsive electrostatic influence that electrons exert on each other. This is not a minor correction: the force between electrons is about as strong as that between electron and nucleus. So for any element aside from hydrogen, Bohr's appealing model becomes too complicated to work out exactly.

In trying to go beyond these limitations, however, Born was not content to fit experimental observations to improvised quantum hypotheses as Bohr had done. Rather, he wanted to calculate the disposition of the electrons using principles akin to those that Isaac Newton used to explain the gravitationally bound solar system. In other words, he sought the *rules* that governed the quantum states that Bohr had adduced.

It became clear to Born that what he began to call a 'quantum mechanics' could not be constructed by minor amendment of classical, Newtonian mechanics. 'One must probably introduce entirely new hypotheses', Heisenberg wrote to Pauli – another former pupil of Sommerfeld in Munich, where the two had become friends – in early 1923. Born agreed, writing that summer that 'not only new assumptions in the usual sense of physical hypotheses will be necessary, but the entire system of concepts of physics must be rebuilt from the ground up'.

That was a call for revolution, and the 'new concepts' that emerged over the next four years amounted to nothing less. Heisenberg began formulating quantum mechanics by writing the energies of the quantum states of an atom as a matrix, a kind of mathematical grid. One could specify, for example, a matrix for the positions of the electrons, and another for their momenta (mass times velocity). Heisenberg's version of quantum theory, devised with Born and Jordan in 1925, became known as matrix mechanics.

It wasn't the only way to set out the problem. From early 1926 the Austrian physicist Erwin Schrödinger, working at the University of Zurich, began to explicate a different form of quantum mechanics based not on matrices but on waves. Schrödinger postulated that all the fundamental properties of a quantum particle such as an electron, or a collection of such particles, can be expressed as an equation describing a wave, called a wavefunction. The obvious question was: a wave of what? The wave itself is a purely mathematical entity, incorporating 'imaginary numbers' derived from the square root of -1 (denoted i), which, as the name implies, cannot correspond to any observable quantity. But if one calculates the square of a wavefunction – that is, if one multiplies this mathematical entity by itself* – then the imaginary numbers go away and only real ones remain, which means that the result may correspond to something concrete that can be measured in the real world. At first Schrödinger thought that the square of the wavefunction produces a mathematical expression describing how the density of the corresponding particle varies from one place to another, rather as the density of air varies through space in a sound wave. That was already weird enough: it meant that quantum particles could be regarded as smeared-out waves, filling space like a gas. But Born – who, to Heisenberg's dismay, was enthusiastic about Schrödinger's rival 'wave mechanics' – argued that the squared wavefunction denoted something even odder: the *probability* of finding the particle at each location in space.

Think about that for a moment. Schrödinger was asserting that the wavefunction says all that can be said about a quantum system. And apparently, all that can be said is not where the particle is, but what the chance is of finding it *here* or *there*. This is not a question of

---

* More strictly, one calculates the so-called complex conjugate, the product of two wavefunctions identical except that the imaginary parts have opposite signs: $+i$ and $-i$.

incomplete knowledge – of knowing that a friend might be at the cinema or the restaurant, but not knowing which. In that case she is one place or another, and you are forced to talk of probabilities just because you lack sufficient information. Schrödinger's wave-based quantum mechanics is different: it insists that *there is no answer* to the question beyond the probabilities. To ask where the particle *really* is has no physical meaning. At least, it doesn't until you look – but that act of looking doesn't then disclose what was previously hidden, it *determines what was previously undecided*.

Whereas Heisenberg's matrix mechanics was a way of formalizing the quantum jumps that Bohr had introduced, Schrödinger's wave mechanics seemed to do away with them entirely. The wavefunction made everything smooth and continuous again. At least, it seemed to. But wasn't that just a piece of legerdemain? When an electron jumps from one atomic orbit to another, the initial and the final states are both described by wavefunctions. But how did one wavefunction change into the other? The theory didn't specify that – you had to put it in by hand. And you still do: there remains no consensus about how to build quantum jumps into quantum theory. All the same, Schrödinger's description has prevailed over Heisenberg's – not because it is more correct, but because it is more convenient and useful. What's more, Heisenberg's quantum matrices were abstract, giving scant purchase to an intuitive understanding, while Schrödinger's wave mechanics offered more sustenance to the imagination.

The probabilistic view of quantum mechanics is famously what disconcerted Einstein. His scepticism eventually isolated him from the evolution of quantum theory and left him unable to contribute further to it. He remained convinced that there was some deeper reality below the probabilities that would rescue the precise certainties of classical physics, restoring a time and a place for everything. This is how it has always been for quantum theory: those who make great, audacious advances prove unable to reconcile them to the still more audacious notions of the next generation. It seems that one's ability to 'suppose' – 'understanding' quantum theory is largely a matter of reconciling ourselves to its counter-intuitive claims – is all too easily exhausted by the demands that the theory makes.

Schrödinger wasn't alone in accepting and even advocating inde-terminacy in the quantum realm. Heisenberg's matrix mechanics seemed to insist on a very strange thing. If you multiply together the matrices describing the position and the momentum of a particle, you get a different result depending on which matrix you put first in the arithmetic. In the classical world the order of multiplication of two quantities is irrelevant: two times three is the same as three times two, and an object's momentum is the same expressed as mass times velocity or velocity times mass. For some pairs of quantum properties, such as position and momentum, that was evidently no longer the case.

This might seem an inconsequential quirk. But Heisenberg discovered that it had the most bizarre corollary, as foreshadowed in the portentous title of the paper he published in March 1927: 'On the perceptual content of quantum-theoretical kinematics and mechanics'. Here he showed that the theory insisted on the impossibility of knowing at any instant the precise position *and* momentum of a quantum particle. As he put it, 'The more precisely we determine the position, the more imprecise is the determination of momentum in this instant, and vice versa.'

This is Heisenberg's uncertainty principle. He sought to offer an intuitive rationalization of it, explaining that one cannot make a measurement on a tiny particle such as an electron without disturbing it in some way. If it were possible to see the particle in a microscope (in fact it is far too small), that would involve bouncing light off it. The more accurately you wish to locate its position, the shorter the wavelength of light you need (crudely speaking, the finer the divisions of the 'ruler' need to be). But as the wavelength of photons gets shorter, their energy increases – that's what Planck had said. And as the energy goes up, the more the particle recoils from the impact of a photon, and so the more you disturb its momentum.

This thought experiment is of some value for grasping the spirit of the uncertainty principle. But it has fostered the misconception that the uncertainty is a result of the limitations of experimentation: you can't look without disturbing. The uncertainty is, however, more fundamental than that: again, it's not that we can't get at the information, but that this information *does not exist*. Heisenberg's uncertainty principle has also become popularly interpreted as imputing fuzziness and imprecision to quantum mechanics. But that's not quite right either. Rather, it places very precise limits on *what we can know*. Those limits, it transpires, are determined by Planck's constant, which is so small that the uncertainty becomes significant only at the tiny scale of subatomic particles.

Political science

Both Schrödinger's wavefunction and Heisenberg's uncertainty principle seemed to be insisting on aspects of quantum theory that verged on the metaphysical. For one thing, they placed bounds on what is knowable. This appeared to throw causality itself – the bedrock of science – into question. Within the blurred margins of quantum phenomena, how can we know what is cause and what is effect? An electron could turn up here, or it could instead be there, with no apparent causal principle motivating those alternatives.

Moreover, the observer now intrudes ineluctably into the previously objective, mechanistic realm of physics. Science purports to pronounce on how the world works. But if the very act of observing it changes the outcome – for example, because it transforms the wavefunction from a probability distribution of situations into one particular situation, commonly called 'collapsing' the wavefunction – then how can one claim to speak about an objective world that exists before we look?

Today it is generally thought that quantum theory offers no obvious reason to doubt causality, at least at the level at which we can study the world, although the precise role of the observer is still being debated. But for the pioneers of quantum theory these questions were profoundly disturbing. Quantum theory worked as a mathematical description, but without any consensus about its interpretation, which seemed to be merely a matter of taste. Many physicists were content with the prescription devised between 1925 and 1927 by Bohr and Heisenberg, who visited the Dane in Copenhagen. Known now as the Copenhagen interpretation, this view of quantum physics demanded that centuries of classical preconceptions be abandoned in favour of a capitulation to the maths. At its most fundamental level, the physical world was unknowable and in some sense indeterminate. The only reality worthy of the description is what we can access experimentally – and that is all that quantum theory prescribes. To look for any deeper description of the world is meaningless. To Einstein and some others, this seemed to be surrendering to ignorance. Beneath the formal and united appearance of the Solvay group in 1927 lies a morass of contradictory and seemingly irreconcilable views.

These debates were not limited to the physicists. If even *they* did not fully understand quantum theory, how much scope there was then for confusion, distortion and misappropriation as they disseminated these ideas to the wider world. Much of the blame for this must be laid at the door of the scientists themselves, including Bohr and Heisenberg, who threw caution to the wind when generalizing the narrow meaning of the Copenhagen interpretation in their public pronouncements. For Bohr, a crucial part of this picture was the notion of *complementarity*, which holds that two apparently contradictory descriptions of a quantum system can both be valid under different observational circumstances. Thus a quantum entity, be it an insubstantial photon or an electron graced with mass, can behave at one time as a particle, at another as a wave. Bohr's notion of complementarity is scarcely a scientific theory at all, but rather, another characteristic expression of the Copenhagen affirmation that 'this is just how things are': it is not that there is some deeper behaviour that sometimes looks 'wave-like' and sometimes 'particle-like', but rather, this duality is an intrinsic aspect of nature. However one feels about Bohr's postulate, there was little justification for his enthusiastic extension of the complementarity principle to biology, law, ethics and religion. Such claims made quantum physics a political matter.

The same is true for Heisenberg's insistence that, via the uncertainty principle, 'the meaninglessness of the causal law is definitely proved'. He tried to persuade philosophers to come to terms with this abolition of determinism and causality, as though this had moreover been established not as an (apparent) corollary of quantum theory but as a general law of nature.

This quasi-mystical perspective on quantum theory that the physicists appeared to encourage was attuned to a growing rejection, during the Weimar era, of what were viewed as the maladies of materialism: commercialism, avarice and the encroachment of technology. Science in general, and physics in particular, were apt to suffer from association with these supposedly degenerate values, making it inferior in the eyes of many intellectuals to the noble aspirations of art and 'higher culture'. While it would be too much to say that an emphasis on the metaphysical aspects of quantum mechanics was cultivated in order to rescue physics from such accusations, that desideratum was not overlooked. Historian Paul Forman has argued that the quantum physicists explicitly accommodated their interpretations to the prevailing social ethos of the age, in which 'the concept – or the mere word – "causality" symbolized all that was odious in the scientific enterprise'. In his 1918 book *Der Untergang des Abendlandes* (*The Decline of the West*), the German philosopher and historian Oswald Spengler more or less equated causality with physics, while making it a concept deserving of scorn and standing in opposition to life itself. Spengler saw in modern physicists' doubts about causality a symptom of what he regarded as the moribund nature of science itself. Here he was thinking not of quantum theory, which was barely beginning to reach the public consciousness at the end of the First World War, but of the probabilistic microscopic theory of matter developed by the Scottish physicist James Clerk Maxwell and the Austrian Ludwig Boltzmann, which had already renounced claims to a precise, deterministic picture of atomic motions.

Spengler's book was read and discussed throughout the German academic world. Einstein and Born knew it, as did many other of the leading physicists, and Forman believes that it fed the impulse to realign modern physics with the spirit of the age, leading theoretical physicists and applied mathematicians to 'denigrat[e] the capacity of their discipline to attain true, or even valuable, knowledge'. They began to speak of science as an essentially spiritual enterprise, unconnected to the demands and depradations of technology but, as Wilhelm Wien put it, arising 'solely from an inner need of the human spirit'. Even Einstein, who deplored the rejection of causality that he saw in many of his colleagues, emphasized the roles of feeling and intuition in science.

In this way the physicists were attempting to reclaim some of the prestige that science had lost to the neo-Romantic spirit of the times. Causality was a casualty. Only once we have 'liberation from the rooted prejudice of absolute causality', said Schrödinger in 1922, would the puzzles of atomic physics be conquered. Bohr even spoke of quantum theory having an 'inherent irrationality'. And as Forman points out, many physicists seemed to accept these notions not with reluctance or pain but with relief and with the expectation that they would be welcomed by the public. He does not see in all this simply an attempt to ingratiate physics to a potentially hostile audience, but rather, an unconscious adaptation to the prevailing culture, made in good faith. When Einstein expressed his reservations about the trend in a 1932 interview with the Irish writer James Gardner Murphy, Murphy responded that even scientists surely 'cannot escape the influence of the milieu in which they live'. And that milieu was anti-causal.

Equally, the fact that both quantum theory and relativity were seen to be provoking crises in physics was consistent with the widespread sense that crises pervaded Weimar culture – economically, politically, intellectually and spiritually. 'The idea of such a crisis of culture', said the French politician Pierre Viénot in 1931, 'belongs today to the solid stock of the common habit of thought in Germany. It is a part of the German mentality.' The applied mathematician Richard von Mises spoke of 'the present crisis in mechanics' in 1921; another mathematician, Hermann Weyl (one of the first scientists openly to question causality) claimed there was a 'crisis in the foundations of mathematics', and even Einstein wrote for a popular audience on 'the present crisis in theoretical physics' in 1922.* One has the impression that these crises were not causing much dismay, but rather, reassured physicists that they were in the same tumultuous flow as the rest of society.

This was, however, a dangerous game. Some outsiders drew the conclusion that quantum mechanics pronounced on free will, and it was only a matter of time before the new physics was being enlisted for political ends. Some even managed to claim that it vindicated the policies of the National Socialists.

Moreover, if physics was being in some sense shaped to propitiate Spenglerism, it risked seeming to endorse also Spengler's central thesis of relativism: that not only art and literature but also science and mathematics are shaped by the culture in which they arise and are invalid and indeed all but incomprehensible outside that culture. It is tempting to find here a presentiment of the 'Aryan physics' propagated by Nazi sympathizers in the 1930s (see Chapter 6), which contrasted

* Experimental physicist Johannes Stark's 1921 book *The Present Crisis in German Physics* used the same trope but spoke to a very different perception: that his kind of physics was being eclipsed by an abstract, degenerate form of theoretical physics – see page 91.

healthy Germanic science with decadent, self-serving Jewish science. And given Spengler's nationalism, rejection of Weimar liberalism, support for authoritarianism and belief in historical destiny, it is no surprise that he was initially lauded by the Nazis, especially Joseph Goebbels, nor that he voted for Hitler in 1932. (Spengler was too much of an intellectual for his advocacy to survive close contact. After meeting Hitler in 1933, he distanced himself from the Nazis' vulgar posturing and anti-Semitism, and was no favourite of the Reich by the time he died in Munich in 1936.)

One way or another, then, by the 1920s physics was becoming freighted with political implications. Without intending it, the physicists themselves had encouraged this. But they hadn't grasped – were perhaps unable to grasp – what it would soon imply.

3 'The beginning of something new'

Maastricht has always been proud that Peter Debye was born there. A bronze bust of the scientist is displayed in the City Hall, and Debye was present when it was unveiled in November 1939, shortly before he left Europe for good. In those troubled times, the (somewhat convoluted) address at the inauguration declared, Debye was 'the loyal soldier and shield-bearer of and at the gate of civilization, [a] master of culture, who preserves the purity of spirit and elevates it to serve, indeed to deserve the Nobel Prize awarded to the great masters among mankind'.

It is ironic that Debye's Dutch nationality eventually precipitated the end of his career in German science, since in many ways he did not seem especially attached to the country of his birth. Rather, he identified more strongly as a *Maastrichtenaar*, a native of the capital city of Limburg, where the confluence of three cultures weakened any sense of national identity. In Maastricht, Debye said, one could use Dutch, Belgian or German currency – a symbolic indication of the city's cultural fluidity.

Peter Debye was born in 1884, the son of a metalworks foreman and a mother who worked as a ticket-seller at the theatre. They lived in a working-class district populated mainly by tradespeople, and until Debye went to school he spoke only the local dialect. It was expected at first that he would enter his father's trade, but his mother had higher ambitions, sending him in spite of the cost to a good secondary school run by monks.

He was an outstanding student in science, and so it was decided that he should go to college – an unusual step in view of his humble origin. University was out of the question, for Debye had not learnt the requisite Greek and Latin, so he went to the technical college at

Peter Debye (1884–1966) at the Kaiser Wilhelm Institute for Physics around 1936.

Aachen to study electrical engineering, which had become a popular subject thanks to the burgeoning electrification of daily life. The feeling was, he later explained, 'that the best thing was for me to become an electrical engineer because I was interested in electricity and had done a lot of experiments myself and built a dynamo machine [at school]'. There was nothing surprising in the decision to continue his studies in Germany: he had already learnt good German at school, and Aachen was only twenty miles away from Maastricht. For electrical engineering, the only viable alternatives were Ypres in Belgium, or Delft. But Aachen was the cheapest option, and that was what mattered to a boy of modest means – it was, Debye said, 'just a question of money'. He lived in a single room shared with another lad from Maastricht, and to further conserve his funds he went home at weekends. On Monday morning he would have to rise in time to take the one o'clock train to Aachen, arriving in time for the eight o'clock lecture despite the one-hour time difference.

Debye enrolled at Aachen in 1901. His abilities soon caught the eye of Arnold Sommerfeld, a young professor of engineering mechanics who had arrived at Aachen just a year before Debye. Sommerfeld found his promising student to be 'a charming boy who looked out on the world and on life with intelligence and curiosity', and he appointed him as his research assistant in 1904. Sommerfeld would invite his best pupils to his house for supper, where they would share a bottle of wine and talk – or for the students, mostly listen. 'We came to his house in the evening at eight o'clock', Debye recalled,

> had the evening meal, supper. And then you sit in his room. And in his room he began to talk. He talked about the things he was interested in, and you sat there as a kind of an audience. He asked you about it, although you did not know anything about it. He tried it out, so to say. And in this way I learned a lot.

They were luckier than probably they knew. Whereas most German academics observed a stern, detached formality, Sommerfeld was generous and respectful, willing even to lend his students a few marks when they ran short. He didn't pretend to be omniscient, but sought his students' help to attack difficult problems. Although his post was in applied science, Sommerfeld's principal interests lay in mathematical physics: he had studied with the mathematicians Ferdinand von Lindemann in his native Königsberg and Felix Klein at Göttingen, and his interests ranged from hydrodynamics to the theory of electrical conduction. At Aachen he lacked colleagues with whom he could discuss these things; Debye suspected that the efforts he put into teaching were partly to cultivate a group of like-minded individuals who could act as sounding boards. He had an eye for talent, but perhaps too a uniquely effective way of nurturing it. Although Sommerfeld's contributions to physics were by no means inconsiderable, his most impressive legacy is his roster of students, which includes Heisenberg, Pauli, Hans Bethe, Max von Laue and Linus Pauling. Even among such glittering company, Sommerfeld apparently considered Debye his greatest discovery.

When Wilhelm Röntgen appointed Sommerfeld head of the new Institute for Theoretical Physics at Munich in 1906, Debye went with him. There he frequented the Hofgarten Café in the gardens of the Royal Palace, where physicists and chemists convened for informal meetings at the *Stammtisch*, the table reserved for regulars. 'Like the Viennese cafés we got the newspapers and we talked about God knows what', Debye recalled. It was here, he claimed, that Max von Laue – sent to Sommerfeld by Planck because he was 'too nervous' for Berlin – realized that X-rays bouncing off layers of atoms would interfere with one another and enable researchers to peer into crystals. Around the table of the Hofgarten, said another of Sommerfeld's students, the Pole Paul Epstein, 'the physicists [would] talk really about what the purpose is of what they are doing and not just the outside appearance, and that would be a way to learn physics'.

Debye also started up a club for the younger members of the faculty. It met every Tuesday between five and seven o'clock to hear a talk from one of them, after which they would retire to a restaurant for supper and then move on to a bowling club. Debye made things happen; he impressed his peers and superiors, and he was popular. According to Paul Ewald, another of Sommerfeld's illustrious students, he was 'even then an outstanding physicist, mathematician, and helpful friend'.

Debye gained his doctorate in Munich in 1908, and three years later, on Sommerfeld's recommendation, he took the seat in theoretical physics vacated by Einstein at the University of Zurich. There was little doubt by this time that he was destined to become a leading scientist. Some already considered him Einstein's intellectual successor, and Einstein himself described Debye in 1920 as 'Sommerfeld's most brilliant student'.

But circumstances, or perhaps a restless nature, led Debye on a wandering course. He was at Zurich for just a year before moving to the University of Utrecht, where he hoped for more opportunity to do experimental work. When that failed to materialize, he moved to Göttingen in 1914, where he worked with the Swiss scientist Paul Scherrer on the new field of X-ray diffraction that Laue had initiated. By 1920 he and Scherrer had transferred back to Zurich, this time to the Federal Institute of Technology (ETH). It was here in 1923, in collaboration with his assistant Erich Hückel, that he developed his important theoretical work on salt solutions, showing how electrical interactions between the dissolved salt ions affect one another's properties.

During his time at Utrecht, Debye returned to Munich to marry

Mathilde Alberer, the daughter of the proprietor of his boarding house there. Their first son Peter was born in 1916 in Göttingen, followed by a daughter, Mathilde Maria ('Maida'), in 1921 in Zurich.

Having completed essentially all of his scientific training in Germany, Debye felt a strong allegiance to German culture. 'I feel myself to be very "German"', he wrote to Sommerfeld in 1912. 'You should not even think that I could betray my German education, because it would be completely impossible for me even if I wanted to.' This affinity for things Germanic, in the cultural rather than the nationalistic sense, was shared by other foreign scientists who worked, as it were, in the shadow of the German tradition. The Dutch physicist Hendrik Casimir, then based at Leiden, averred that

> A good deal of German culture had gone into my idiosyncrasy: I had learned much from German books, German was the first foreign language I spoke fluently, my father had been strongly influenced by the German philosophers.

He added revealingly that this 'made it difficult for me to identify Germany – even Nazi Germany – with the Devil'. Despite this regard for German culture, however, Debye evidently had no intention of ever becoming a German citizen. And his wife, by marrying him, forfeited her German citizenship and became Dutch.

By the time Debye was appointed professor of experimental physics at the University of Leipzig in 1927, he had completed his major contributions to science. None of these had the iconoclastic aspect of relativity or quantum theory, and it isn't easy to convey their significance or even their content to non-scientists. Yet Debye's discoveries into the way light, electrical forces and molecules interact had immediate practical consequences in areas ranging from the design of batteries to the understanding of how liquids are structured. According to his biographer Mansel Davies, it is debatable 'whether, in the broad area of molecular physics, any single individual since Faraday has contributed so much'. Since it is also debatable whether there is a less fashionable field of science than molecular physics, Debye's scientific eminence has remained more or less unknown in the wider world. And yet he came to Leipzig as one of the major players in German physics, a theorist with an experimentalist's acuity. And he seemed to have ascended to this height almost casually, to the chagrin of some of his peers. 'Clever but lazy' is how his fellow faculty member Friedrich Hund described him, saying that he would often spot Debye smoking a cigar or watering the roses in the institute garden when he should have been working. 'Debye', he concluded, 'had a certain tendency to take things easy.'

The idealist

Shortly after Debye arrived at Leipzig, and largely at his recommendation, the university appointed a new professor of theoretical physics. Aged just twenty-six, Werner Heisenberg was the youngest incumbent of a physics professorship in all of Germany.

Brought up in an austere, well-to-do and militaristic family in southern Germany, Heisenberg experienced little by way of emotional development or imaginative stimulation. As a result, he seems to have craved comradeship while possessing few of the characteristics that make it easy to come by. On the contrary, he was nervous, competitive, desperate for recognition, and inclined to withdraw, or occasionally to become angry, if slighted or bettered.

Where Debye was able to penetrate swiftly to the physical core of a problem, Heisenberg tended to work in the other direction: from the concrete to the abstract. He was a gifted mathematician who could shape profound questions into formalized terms. For Heisenberg abstraction seemed to offer a degree of philosophical satisfaction, even consolation. He was arguably the most gifted physicist of his generation, an idealist to Debye's pragmatist. It is with the nature of those ideals that historians have wrestled and argued, and continue to do so.

Heisenberg spent his formative student years with Sommerfeld in Munich, where he arrived in 1920 amidst the political unrest that culminated in the failed putsch by the Nazi Party. These upheavals left Heisenberg with a disdain for politics, and he sought refuge in the youth movement called the New Pathfinders. In a reverence for nature, music and the Romantic German writers such as Goethe and Hölderlin, Heisenberg found a world that transcended the power struggles of the Bavarian Communists and National Socialists. It was the same release that he sought in physics.

Werner Heisenberg (1901–1976) in 1927.

Heisenberg remained a youth leader well into his adult life; in these mountain hikes he assuaged a hankering after youth that began to seem increasingly peculiar as he grew older. When he married Eliza-beth Schumacher, sister of the economist E. F. Schumacher, in 1937, he was thirty-five and she just twenty-two. This preference to be with people much younger than himself has been adduced as an indication of Heisenberg's latent immaturity, although such an age disparity with their spouse was not uncommon for German academics, who often had to wait many years for a position that conveyed marriageable stability. Despite the solace Heisenberg drew from his marriage, the relationship was in some ways never a close one – Elizabeth admitted later that she and her husband never really got to know one another, and Heisenberg played very little part in bringing up his children. His somewhat awkward, occasionally insensitive and bristly persona was the outward manifestation of inner turmoil, anxiety, and a sense of loneliness and alienation, yet he was rarely able to confide any of this to his wife. Both his ambition and his craving for approval speak of a deep insecurity.

The New Pathfinders yearned for a powerful leader who would deliver Germany from Weimar decadence and restore national honour. This did not mean that Heisenberg welcomed Hitler's ascendancy, but it made him, like many Germans from upper-middle-class families, favourably disposed towards some aspects of the National Socialists' policies, not least their militaristic truculence. 'The beginning of something new and "solid" is not yet upon us', he wrote to his father in 1922 with breathless anticipation. While it would be too much to regard the New Pathfinders movement as a precursor to the Hitler Youth, it was paramilitary, puritanical and nationalistic, and when such independent youth groups were outlawed in 1934, many members found continuity in transferring to the Nazi youth organization.

During his Munich years with Sommerfeld, Heisenberg worked on the difficult but decidedly classical physics of turbulence in fluid flow. But Sommerfeld also conscripted his bright student in the quest to place quantum theory on a mathematical footing. Heisenberg's success in describing certain aspects of the Zeeman effect – the influence of magnetic fields on light emission from atoms (page 24) – was related by Sommerfeld to Einstein in an enthusiastic letter at the start of 1922. When Heisenberg left Munich for Göttingen to complete his Habilitation (the postdoctoral qualification that entitles German researchers to hold a faculty post and supervise doctoral candidates) with Max Born, his heart was set on formalizing quantum theory. It was widely accepted that the matrix mechanics that he, Born and Pascual Jordan devised turned the ad hoc, qualitative picture of early quantum theory into a robust predictive science. In May 1926 Heisenberg arrived in Copenhagen as a university lecturer and assistant to Niels Bohr; over the following year they devised the Copenhagen interpretation, and Heisenberg revealed the irreducible uncertainty lurking in his matrices.

He was, then, quite a catch for Leipzig. That was amply demonstrated three years after the appointment, when Heisenberg was awarded the Nobel Prize in Physics. Ambitious though he was, he had the grace to be embarrassed by the award citation – 'for the creation of quantum physics' – for he made no pretence of having done anything quite so grand single-handedly. True, Schrödinger was rewarded the next year, but only in conjunction with the British physicist Paul Dirac. Other pioneers such as Wolfgang Pauli had to wait for over a decade more before their contributions were recognized by the Nobel Committee. But even though Heisenberg's diffidence was genuine, the award cannot but have contributed to a growing conviction that he carried the fate of German physics on his shoulders.

4 'Intellectual freedom is a thing of the past'

'Last week', physicist Lise Meitner wrote from the Kaiser Wilhelm Institute for Chemistry in Berlin in March 1933 to her colleague Otto Hahn on a visit to North America, 'we received instructions to raise the swastika flag next to the black-white-red one.'

Adolf Hitler had been appointed Reich chancellor only at the end of January of that year, but already Germany was becoming a dictatorship. The transition from democracy to totalitarian regime was frighteningly swift, as Hitler moved immediately to eliminate political opposition. An arson attack on the Reichstag building on 27 February was blamed on Communist agitators, giving Hitler an excuse to declare a state of emergency, suspend civil liberties and impose press censorship. By fomenting general panic among the population, the Nazis also orchestrated justification for the Enabling Law 'to remedy the distress of the people and the state'. When it was passed in March, the law gave Hitler power to legislate without the consent of the Reichstag and even to overrule the constitution. The Civil Service Law followed on 7 April, expelling not only Jews but also all potential opponents from places of power and influence. As historian Ian Kershaw says, 'in 1933 the barriers to state-sanctioned measures of gross inhumanity were removed almost overnight. What had previously been unthinkable suddenly became feasible.'

Yet to many Germans these were the actions not of a ruthless dictator but of a resolute leader: they were a necessary means of bringing about public order and security in the face of the dangers created by the Weimar state. The plummet into total rule was accompanied by a widespread sense of relief and optimism. At last the political chaos seemed to have been tamed by a decisive and strong Führer who reflected the prevailing mood of conservative nationalism.

Even Meitner, an Austrian Jew, did not view the Nazi government with outright foreboding. She told Hahn that Hitler had recently spoken 'very moderately, tactfully, and personally'. Hopefully, she added, 'things will continue in this vein'. For his part, Hahn expressed some support for Hitler's cause while abroad, telling Canadian reporters he was under the impression that all the Jews who had been ousted and incarcerated were Communist agitators.

The democratic voice

Hitler understood that he could do pretty much as he pleased so long as he made it legal. By introducing their discriminatory and authoritarian policies in the form of new laws, the National Socialists exploited the German instinct for obedience to the state: one did not object to measures that were enshrined in law. The Prussian mentality in particular, trained to be loyal and subservient to authority, found it almost inconceivable to oppose what was state-ordained. The idea that laws could be morally criminal was virtually a contradiction in terms.

All the same, how a conservative but humane society could capitulate to leaders with such vicious and abhorrent objectives has been the central question for historians of the Nazi era. The first consideration is obvious yet easily forgotten: we know now where Germany was heading, but the German citizens did not. It seems trivial to say that, to understand the events of 1933, we must set aside vision of the impending Holocaust. But it is hard to do so. As Alan Beyerchen says, 'only in retrospect is it so apparent that the only truly honourable response to National Socialism was uncompromising defiance'.

We should remember too that modern European society, while no stranger to bigotry and undemocratic rule, had no previous experience of this sort of extreme demagoguery, state repression and legalized racism. It was widely believed that Hitler's government would be transient: that it would either soon lose power or be forced to moderate its extremes. That seemed to be the impression gathered by Wilbur Earle Tisdale of the Rockefeller Foundation's European office on his excursions among German scientists: he wrote to his New York superiors in August 1934 that 'observers inside and outside of Germany [are] unanimous in predicating [sic] the fall of the present government within a period of a few months'. He reported that Otto Warburg, director of the Kaiser Wilhelm Institute of Cell Physiology in Berlin, suspected that the Nazis in fact ultimately planned to restore the monarchy!

Yet we cannot ascribe the ascent of the Nazis simply to over-optimism and lack of foresight on the part of the German people. Hitler's party mobilized pervasive patriotism and deep-seated (if often mildly expressed) ethnic prejudice, coupled to the general population's parochial self-interest, political and economic dissatisfaction, fear of Bolshevik revolution, and instinct to avoid trouble. These tendencies could be found across the political spectrum. Watching the options for political opposition evaporate, the despairing chairman of the moderate right German People's Party (DVP) wrote that 'behind the pretty facade of patriotic unity, do not infinitely many people operate only out of ambition, greed, class hostilities, and a desire for advance-ment to a degree that endangers the personal trust between ordinary Germans in the worst way?'

Besides, in many respects the National Socialists did not look like far-right fanatics at all – why then would they be styling themselves as socialists? Their welfare policies seemed progressive; their economic strategy was Keynesian. And after the turmoil of the Weimar Republic, people of all political persuasions could see the attraction of firm, conservative government. 'From 1929 on,' wrote the civil servant Hans Bernd Gisevius,

> it became more and more apparent that the leaders of our left and centre parties were incapable of holding the masses in line. It seemed quite reasonable to hope that the rising flood [of public unrest] could be stemmed by the right and safely guided into evolutionary channels.

Gisevius, whose instincts were fundamentally those of the conservative Prussian elite, felt that German liberalism* 'must bear a considerable

* The question of what exactly German liberalism *was* in the 1930s is not straight-forward, since political lines were not drawn in the same way as they tend to be in Western democracies today. Political culture was largely polarized between right and left, in which camps one might identify moderates and extremists – there was scarcely a 'political middle' at all. The Weimar government was formed initially from a coali-tion of the Social Democratic Party of Germany (SPD), the German Democratic

measure of guilt for the disaster of Nazism'. With its 'overemphasis on individualism', he said, it 'contributed greatly to the dissolution of religious and ethical principles'.

Although this comment tells us something about why Gisevius and his ilk initially welcomed the Nazis (he applied to join the party in 1933, hoping to gain professional advantage from it, but was refused), it nonetheless places too much blame on liberalism for provoking its antithesis. The fact was that liberals were not necessarily opposed to Hitler in the first place. 'Even though liberal democrats disagreed with National Socialism on some levels', writes historian Eric Kurlander, 'they exhibited indifference, even enthusiasm for the regime on others . . . When liberals failed to resist, at least intellectually, it had less to do with fear of arrest or persecution and more to do with a tacit desire to accommodate specific policies.' The National Socialists were offering much more than what they have now come to represent: state oppression, torture, racism and genocide.

Werner Heisenberg fits this picture of the 'conservative liberal' optimist: in October 1933 he wrote that 'much that is good is now also being tried, and one should recognize good intentions'. His close colleague Carl von Weizsäcker told writer Robert Jungk in the 1950s that (as Jungk put it) 'although he [Weizsäcker] had a loathing for the leaders of this "movement" . . . in its beginnings he had a certain sympathy or, let's say, understanding for National Socialism, because it appeared to him that there was the thrust of profound forces operating here'. That notion of 'profound forces' appears repeatedly in the attitude of the German scientists: a sense that politics were directed by tectonic influences they could not and should not hope to influence – and which were almost by definition of noble character, even if their immediate manifestations were squalid.

Liberals and conservatives united in supporting Nazi foreign policy. Even the Nobel Peace Prize laureate and democrat Ludwig Quidde,

---

Party (DDP), and the German Centre Party – although this coalition lost its majority in the 1920 elections, leaving the democratic voice fatally weak. But while the DDP in particular was the only true 'left liberal' party, there was also a liberal element on the political right, represented by the German People's Party (DVP).

I shall use 'liberal' here to refer to individuals inclined towards democratic ideas, while recognizing that this might encompass quite distinct views on a range of political issues. That, indeed, is the point.

who was exiled from Germany in 1933 and criticized the National Socialists thereafter, approved of rearmament and of Hitler's plans to reunite Austria with Germany, claiming that this is what the Austrians wanted (many of them did indeed welcome the *Anschluss* in 1938). One could find liberals who defended the invasion of Poland; even the invasion of France found favour with some. They regarded these acts as part of a struggle for German liberation.

It is in the light of such considerations that the German historian Karl Dietrich Bracher writes – with the German scientists specifically in mind – that

> Certainly opportunism and fear at the inception of the terrorism of the new regime were contributing factors [to the rise of Nazism]. But the coordination and self-identification with the totalitarian regime occurred to such an astonishing extent and with such speed that one cannot avoid the conclusion that there existed on the part of the great majority of the intellectual elite a very high predisposition and susceptibility towards it.

When liberals did disagree with the regime, they were not necessarily prevented from saying so. The newspaper *Frankfurter Zeitung* published many articles critical of Nazi policies, but was not shut down until 1943, and was hitherto treated with remarkable lenience by the propaganda minister Joseph Goebbels, who understood the value of appearing to tolerate dissent. One can also overestimate the dangers of protest, at least in the early years of Nazi rule. The threat of deportation to the camps or of a Gestapo interrogation cell did not necessarily loom. When, for example, the liberal writer, intellectual and politician Theodor Heuss refused to supply verification of his Aryan ethnicity, he suffered nothing worse than being denied membership of the Reich Association of the German Press. Without his press pass, he was stripped of the editorship of his political journal *Die Hilfe* in 1936, although he regained a pass the following year. In 1934 *Die Hilfe* published, without serious consequences, an anti-censorship article arguing that 'religion, science and art are not means to be employed by the state; for the intellect blows wherever it wishes'. One would struggle to find comparable public statements from German scientists.

The National Socialists were in any case insufficiently organized to suppress all criticism. Their apparent ruthlessness and focus in the early days of the Third Reich can give a misleading impression of how the regime mostly worked. It was no monolithic behemoth but was manipulated by rival power blocs, riven by internal factionalism and hindered by bureaucracy and incompetence. Besides, even if it had been a paradigm of efficiency, the secret police (Gestapo), numbering just 20,000 in 1939, was incapable of closely monitoring a population of eighty million. Neither did the Nazis seem particularly bothered to do so: in the pre-war years particularly, they did not 'mould' German mass society by force. 'On close examination', says Kurlander,

> we can find in the [early] Third Reich elements we would not expect in the dictionary definition of a totalitarian regime: a lack of control-ling mechanisms, creative movements expressive of freedom such as jazz and swing, and extended influence of Jewish culture and its cham-pions, even avant-garde attempts at modernism.

Nor was the regime implacable in its resolve: some Nazi policies were revised in the face of public discontent, such as the 'euthanasia' programme that permitted extermination of disabled people.

The fact is that the National Socialists didn't need always to show an iron fist. They seemed to realize that their opponents among the intellectuals would lack the stamina and resources, and probably the convictions, to pose any real threat. The political scientist and historian Alfred Weber at the University of Heidelberg resisted crude Nazi propagandizing in 1933 by insisting that the local police chief remove the swastika banner from his institute and then, when it was restored, simply closing the institute down. Although Weber suffered little more than being pilloried in the Nazi press for this 'suppression of academic freedom', he subsequently elected to retire and pursue his noncon-formism with academic privileges and reputation intact. As Kurlander puts it,

> Weber's abrupt change of heart indicates how daunting it must have been for even the most principled and influential academics to sustain their opposition in the face of goose-stepping students, cowardly

administrators, and career-minded colleagues. Weber's two-week refusal to coordinate [his institute] is the exception; his passive intellectual nonconformity over the ensuing twelve years the rule.

As with some other modern dictatorships, the Nazis realized that strength comes not from brutal repression but from winning support with propaganda and populism, including the creation of a leadership cult. Time and again one finds Hitler disassociated in the minds of many Germans from the ugly acts of his underlings: whatever the failings of the Nazi Party, the Führer himself retained his popular appeal as a symbol of national pride, hope and regeneration. Even (perhaps especially) Nazi sympathizers clung to this vision of a flawless leader presiding over bungling and infighting bureaucrats.

When the German press did carry complaints about the regime, these were typically expressed as economic rather than political dissatisfactions. People might not trouble themselves too much about moral principles or academic freedom, but they cared about food. When the economic situation deteriorated in 1935–6, leading to food shortages, there was unrest among the working class. But many of these grumblers would be found eagerly proclaiming their support at the next Nazi rally. 'Rather Hitler than Stalin' was the prevailing view.

All this should prevent us from attributing too much specialness to the rise of the Nazis. While it surely involved a great deal of historical contingency, and while the determination of some historians to see Nazism as a uniquely German phenomenon warrants consideration, insisting too stridently on a Teutonic obsession with 'blood and culture' not only recycles the Third Reich's own baleful tropes but negates the useful lessons we might draw from it. 'The Third Reich was the product of a liberal democracy not unlike [the] contemporary United States', says Kurlander, who adds that our susceptibility to such regimes becomes particularly strong in times of economic and political unrest, just as the rise of far-right parties in Europe today clings to the coattails of recession. It is not fanciful to draw a parallel between the extreme polarization of political culture in Weimar Germany, which made the country all but ungovernable, and recent political trends in the United States. At any event, the question of how German intellectual, scientific and academic cultures fared in the 1930s is not devoid of contemporary relevance.

The plain fact is, then, that Nazism was not imposed but accepted and even welcomed. From Leipzig in May 1933, the author Erich Ebermayer* wrote

> One becomes ever more lonely. Everywhere friends declare their faith in Adolf Hitler. It is as if an airless stratum surrounds us few who remain unable to make such avowals. Of my young friends it is the best who now radically proclaim their allegiance to National Socialism . . . They run around in the plain Hitler Youth uniform, radiant with happiness and pride.

The Jewish question

No one was under any illusions about the anti-Semitism of the Nazi Party. That was clear enough from its leader's early manifesto *Mein Kampf*:

> The Jewish people, with all its apparent intellectual qualities, is nevertheless without any true culture, especially without a culture of its own. For the sham culture which the Jew possesses today is the property of other peoples, and is mostly spoiled in his hands . . . the personification of the Devil as the symbol of all evil assumes the living shape of the Jew.

Of course, anti-Semitism was already deeply ingrained in the culture of the German-speaking nations. Gustav Mahler's travails as director of the Vienna Court Opera from 1897 to 1907 are but one well-known example: even after converting to Catholicism (as many of Jewish origin did), he was relentlessly attacked by the press on ethnic grounds. Vienna was particularly nasty, but there was little there that would raise eyebrows throughout Germany. Anti-Semitism was more than popular prejudice, for it found some endorsement at the 'highest'

---

* Ebermayer was no anti-Nazi dissenter: his case was far more ambiguous. A homosexual and something of a hack who could write fast and with versatility, he appeared to dislike the National Socialists, and helped to hide his Jewish secretary. Yet some of his fiction and screenplays written during the Nazi era were approved by the authorities, and Goebbels in particular defended him – a favour that Ebermayer repaid after the war by co-authoring a biography of Goebbels called *Evil Genius*.

intellectual and political levels. Immanuel Kant had advocated the 'euthanasia of Jewry', albeit by a conversion to 'pure moral religion stripped of all laws and rituals'. In the late nineteenth century the Berlin-based historian, politician and philosopher Heinrich von Treitschke, a member of the Reichstag, publicly supported anti-Jewish sentiment and accused the German Jews of failing to assimilate. His statement in 1879 that 'The Jews are our misfortune!' became the motto of the newspaper *Der Stürmer*, the mouthpiece of Nazi anti-Semitism. (While Treitschke did not actually voice this opinion, merely writing that it was one heard all over Germany, it's not clear that he would have disagreed with it.)

To both Jews and non-Jews in Germany, Nazi anti-Semitism must have seemed, at least initially, just more of the same. As the Jewish writer Joseph Roth commented bitterly, writing in exile from Paris in the autumn of 1933,

> If you want to understand the burning of the books, you must understand that the current Third Reich is a logical extension of the Prussian empire of Bismarck and the Hohenzollerns, and not any sort of reaction to the poor German republic with its feeble German Democrats and Social Democrats.

Even shortly after the war, when the realities of the Final Solution had become known, one survey indicated that about one in five Germans agreed with Hitler's treatment of the Jews and another one in five were broadly in favour but felt that he went too far.

Germany was not alone in any of this, as the Dreyfus affair in France illustrated. But although distrust of the Jews was widespread in Europe, it didn't have the same flavour everywhere. We are apt now to make the assumption that anti-Semitism was a clearly defined and uniform position associated with the conservative right, predicated on a general disparagement of non-European races, and opposed (when at all) for moral reasons. But none of these things is generally true.

One could, for example, identify with many liberal beliefs while indulging pronounced, even virulent anti-Semitism. As Kurlander says, 'the aspects of Nazi ideology that most offend modern liberals – its virulent, expansionist *völkisch* nationalism and racial anti-Semitism – were the least problematic components of National Socialism for a great number of democrats during the last years of the Weimar Republic'. There was no social stigma, no self-censorship, that might restrain one from casually expressing anti-Semitic sentiments, any more than other forms of prejudice and stereotyping.

Yet German liberalism, more than that of other European nations, found particular room for dislike of Jews. Especially after the First World War, there was widespread doubt in Germany that ethnic minorities could ever be assimilated – or at the very least, assimilation was considered the price of social acceptance. The German political writer Paul Rohrbach provides a good example of how such attitudes were expressed among intellectuals. Having travelled extensively in Asia and Africa, he had a high regard for Chinese and Japanese culture, and was optimistic about the development prospects of Africa. He held socially progressive views, particularly concerning the status of women. Yet he was a strong patriot, convinced of the need to preserve national honour, and bigoted about race. In 1933 he joined the NSDAP, although his relationship with the party was often fraught.

The 'Jewish question' was regarded as a matter of politics, not morality. One might debate it in much the same spirit as one discussed the conduct of trade, war or taxation. Like racism today, it could be seen as nothing personal: you could lament an excessive Jewish influence on politics and commerce, or perpetuate anti-Semitic stereotypes, while enjoying good friendships with Jews. Hjalmar Schacht, Hitler's economic minister, offers an interesting case study of the complexities of anti-Semitism at this time. He was in many respects a liberal, and although he became a supporter of the Nazis and president of the Reichsbank, he lost his influence after a disagreement with Hitler in 1937 ('You simply do not conform to the general National Socialist framework', Hitler told him two years later) and eventually became a member of the German Resistance. He was imprisoned after the failed assassination plot of June 1944 and sent to Dachau, but survived. Schacht was instinctively averse to racial hatred, and was frequently reprimanded by party officials for speaking out against attacks on Jews and their property. He argued against some anti-Semitic measures on the grounds that they would weaken Germany domestically and isolate it abroad. Put on trial at Nuremberg, Schacht claimed that he had served in the government 'to prevent the worst excesses of Hitler's policies', although some historians argue that he aided the Holocaust by expropriating Jewish property. He was acquitted at the trials, and later became an adviser to developing countries on economic development. Schacht's trajectory shows how unwise it is to attempt to label individuals as Nazi or not, or as pro-/anti-Semite. We can find liberals quietly acquiescing to a pervasive anti-Semitism, and on the other hand some prejudiced people driven by the Nazi excesses into defending the Jews. Few scientists actually served in the Nazi administration; but few, too, spoke out publicly against the regime and actively opposed it. Does this make them better or worse than Schacht?

The lack of resistance to the Jewish persecutions in and after 1933 does not necessarily indicate acceptance, but rather, in historian Ian Kershaw's chilling phrase, 'lethal indifference'. There is plenty of indication that ordinary Germans found naked displays of brutality abhorrent, but as a political and moral issue the 'Jewish question' simply did not seem to have much relevance for their daily lives. Sustaining this indifference, however, had to be an increasingly active decision: it meant turning one's back, telling oneself that one was not personally responsible and was in any case powerless to do anything. 'Self-preservation is not a particularly admirable instinct', says Kershaw, 'but especially in a climate of repression and terror it is usually stronger than the instinct to preserve others. It goes hand in hand with moral indifference and apathetic compliance.' And such indifference and compliance, Kershaw believes, are common enough in liberal democracies, let alone in dictatorships. This attitude also captures something of the flavour of the scientists' response to Nazism generally: it is not ideal, but it is none of our business so long as we can avoid its drawbacks. Why go looking for trouble?

Perhaps more surprising than any of this is the fact that Hitler's role in the anti-Semitism that followed in the wake of the Nazi takeover was not immediately clear. Visiting scientists in Berlin in May 1933, Warren Weaver of the Rockefeller Foundation commented that the campaign of Jewish oppression 'is the result of a very deep and general feeling on the part of the common people, and that the government is moderating rather than stimulating the campaign'. Hitler himself, said Weaver, is widely considered 'an influence for moderation'.* He suggests that the Nazis were even a little taken aback by the depth of feeling that they had unleashed: 'They are almost frightened by their own acts, and do not know which way to turn.' Weaver adds that many Germans believed that Jews were agitators for Communism, while others felt that the Nazi anti-Semitism was not ideologically motivated at all but was a diversionary tactic to mask the country's economic woes. Weaver was told that 'the new government has to give the people something, could not give them work, so gave them the Jews instead'. That ruse, he said, 'will not satisfy the crowd for long'.

The purge

The first of the National Socialists' official anti-Semitic measures was presented as an attempt to reduce what many Germans regarded as the unhealthy control that 'the Jews' exerted on the nation's commerce, culture and administration. The Civil Service Law of 7 April 1933 stipulated that 'Civil servants who are not of Aryan descent are to be placed in retirement; in the case of honorary officials, they are to be dismissed from office.' Such appointments included all university posts.

To be an 'Aryan', one needed to prove that both parents *and* grandparents were of Aryan stock – that is, non-Jewish.† People who had

---

* This disassociation of Hitler from anti-Semitic extremism finds a particularly poignant expression in the diaries of the Jewish historian Willy Cohn, who could write even in October 1939, after listening to one of Hitler's speeches, that it was moderate and 'not particularly anti-Semitic', and that 'one should acknowledge the greatness of the man who has given the world a new face'. Like many German Jews, Cohn considered himself more German than Jewish. In 1941 this patriot, a veteran of the First World War with an Iron Cross, was deported with his wife and two daughters to Kaunas in Lithuania, where they were shot along with 2,000 other Jews.
† The 1935 Nuremberg Laws ('for the Protection of German Blood and German Honour') modified this requirement in an attempt to specify the distinctions more clearly: 'Jews' were those who had at least three Jewish grandparents, or two if the person was a practising Jew or was married to one. This meant that people who were formally a quarter Jewish were exempt. The 1935 laws also forbade intermarriage or sexual relations between Jews and Aryans. As these laws indicate, the whole notion of 'Jewishness' is complicated in the Nazi era, being thrust on individuals who felt little affinity for the culture and beliefs of their Jewish forebears. Historian Stefan Wolff argues that using the term 'Jew' in this period risks acquiescing to the Nazi racial ideology unless it is cordoned between quote marks.

shown their loyalty to the nation by serving at the front in the First World War were exempt from the ruling – or at least, that was the principle. In practice these exceptions were subject to the whims of the officials. Warren Weaver describes one veteran scientist, wounded at the front, who was given the impossible task of proving his war record within twenty-four hours to avoid being sacked. 'The world-renowned intellectual freedom of Germany', Weaver concluded, 'is a thing of the past.'

These measures seem unambiguously pernicious today. In Germany in the 1930s that was by no means clear. Some felt that formalizing the principles dividing 'true Germans' from 'Jews' could establish a basis on which they could coexist, and would restrain some of the excesses of anti-Semitic sentiment that threatened to erupt in German society. Even those who did speak out against Jewish exclusion were mostly careful to emphasize that it was the principle of discrimination that they contested, rather than that they were favourably disposed towards Jews. They typically sought to defend Jewish friends and colleagues, not some universal human right of Jewry. They did not question the distinction between 'foreign' Jews and the German *Volk*, nor challenge the *völkisch* ideology, but merely felt that this was an unseemly way to treat people. Even that feeling waned as the anti-Semitic laws made Jews ever less evident in daily life: their very invisibility helped to promote the lethal indifference that permitted Auschwitz.

The Civil Service Law fell particularly hard on German physicists, since around one quarter of them in 1933 – and many of the most able – were officially 'non-Aryan'. This situation was more acute than for the other sciences because physics, being a relatively new subject, had been less afflicted with the prejudices that militated against the advancement of those with Jewish heritage in more conservative and traditional disciplines.

Among those who faced exclusion by the anti-Semitic law were Albert Einstein, Max Born, Eugene Wigner, James Franck, Hans Bethe, Felix Bloch, Otto Stern, Rudolf Peierls, Lise Meitner and Samuel Goudsmit. Many of these names are now attached to physical laws and principles, institutes, awards, chemical elements: it is a roster of Germanic pre-eminence in mid-twentieth-century physics. Some of those affected left the country at once; Einstein was in the United

States when Hitler came to power, and vowed not to return. Others, like Peierls, had seen which way the wind was blowing and already taken posts abroad. One or two, like Meitner, managed to stay for several years more, at increasing personal risk. No prominent 'non-Jewish' scientists quit Germany in protest at their colleagues' plight, however. Erwin Schrödinger made no secret of his dislike of Nazi anti-Semitism when he left for Britain in 1933; but his wife was 'non-Aryan'.

German theoretical physics was decimated. At the University of Göttingen, a major centre for this young discipline, a quarter of the faculty was lost. Often the dismissals were imposed in the most offhand and brutal way. The biochemist Hans Krebs at Freiburg, a student of Otto Warburg, was told without a moment's warning to get out of the laboratory and never set foot in it again. 'I don't think he had time to pick up his handkerchief', Warburg related. Krebs went to Cambridge, and won a Nobel Prize twenty years later.

The response of the German scientific community to these edicts looks today disturbingly compliant. Even a cynic would have to doubt that a governmental ban on racial minorities in the university posts of a modern European country would fail to incite mass protests and resignations. But in Germany in 1933 there was little more than private expressions of dismay, almost always with reference to the harm that was being done to German intellectual status and international reputation rather than to moral values. This is not to make some fatuous point about how much more 'principled' society has now become, but to illustrate how different German society was in the 1930s, and how differently even liberal intellectuals conceptualized and compartmentalized their place in that society.

But wouldn't fear of reprisal, of arrest and the concentration camps, have understandably silenced protest? Certainly it is easy to demand bravery of others, especially in another time and place. However, while the Nazis had a thuggish reputation even before they became the ruling party, the Third Reich was not like Stalinist Russia – not in 1933, and not really subsequently. As we saw earlier, moderately criticizing the regime through official or popular channels might not even do much harm to one's career. Max von Laue did nothing to disguise his anti-Nazi sympathies, but he remained a professor at the University of Berlin until 1943 and the assistant director of the Kaiser

Wilhelm Institute for Physics throughout the war. The personal dangers of defying Hitler may, at least before the war began, have been more imagined than actual.

A stronger deterrent seems to have been the conviction that protest should be seemly and respect protocol. If official appeals were ignored, further resistance was deemed both improper and futile. Max Planck was representative of, and instrumental to, this attitude. According to John Heilbron, he was 'temperamentally unfit for public protests against constituted authority'. For a man like Planck, dedicated to the service of the state and homeland, open defiance was unthinkable. If the government could not be dissuaded from its course in a decorous manner, then there was nothing else to be done. It was precisely because Planck was regarded with such respect, and was seen as embodying the untainted spirit of German science, that his actions were regarded as exemplary: Planck would know what to do. And so whatever Planck did was the proper thing.

And what was that? Planck's initial response to the dismissals reflected a common perception that they were nothing to be too concerned about, that this burst of anti-Semitism would relieve existing tensions and soon the situation would settle into a more tolerable state. He was on holiday in Sicily when the news broke, and at first he saw no pressing need to return and deal with the implications for the KWG, of which he was president. To those who were more worried than he, Planck suggested that they too take a break abroad – by the time you return, he told them, 'all the troubles will be gone'. This looks now like indifference, but it is more properly seen as a grave misapprehension of the nature of the National Socialists – a fatal inability to smell their corruption, allied no doubt to a naïve faith that one's leaders will ultimately see sense. It has been claimed that when Otto Hahn asked Planck whether there should not be some protest on behalf of those who had lost their jobs, he replied that this would be pointless. 'If 30 professors appealed the new measures', he said, '150 would counter them because they wanted the new positions.'*

Planck's complacency was shared by Heisenberg. Hearing of the intention of Max Born (a Lutheran with Jewish roots) to resign at

---

* This apparently telling remark was reported only after the war, so its authenticity has been questioned.

Göttingen, Heisenberg wrote to him in June imploring him to stay for the sake of German physics. He told Born that Planck had 'received the assurance that the government will do nothing beyond the new Civil Service Law that could hurt our science' – in other words, it would get no worse than this. 'In the course of time', he assured Born, 'the splendid things will separate from the hateful.'*

Even Born himself exhibited misplaced optimism. Despite Heisenberg's appeals, he did leave Germany in 1933 to go to Cambridge in England. But on a return visit three years later he was impressed by the efficiency of the 'labour camps' where well-fed and happy men seemed better off than they were while unemployed during the Weimar Republic. While Born was neither blind to nor uncritical of the virulent anti-Semitism he saw on that occasion, his remarks show

---

* A strange story has circulated about Heisenberg and his mentor Born. In his 1985 book *The Griffin*, the Los Alamos physicist Arnold Kramish claims that an anonymous associate of Born's told him of a return visit that the exiled Born had made to Heisenberg in Göttingen some time around 1934, when he was subjected to 'anti-Jewish sneers and obscenities', culminating with Heisenberg spitting at Born's feet. Kramish's confidant said that Born had confessed this incident very reluctantly on his return to England, and that his wife subsequently admitted that it had reduced Born to tears. Since this story places Heisenberg in such an exceedingly bad light, one must demand good evidence of its veracity. Kramish, whose book demonizes Heisenberg relentlessly, gives none.

The episode is repeated by historian Paul Lawrence Rose in his book *Heisenberg and the Nazi Atomic Bomb Project* (1998), which is far more scholarly than Kramish's but equally critical of Heisenberg. Rose, however, claims that Kramish was wrong about the date of the affair, saying that it happened instead in 1953. He thanks Kramish for privately providing details, but he too gives no further documentation. Not only is there no mention of the allegations in Born's autobiography or in any other accounts of Heisenberg's life, but it is hard to see how it could be consistent with the lifelong friendship that the two men apparently sustained. Physicist Frederick Seitz rightly says that the story 'do[es] not appear to fit in with the true relationship between the two scientists'. Seitz also refers to a letter from Born's son Gustav to a friend who, after seeing Kramish's book, had written to ask about the incident. 'I have serious doubts about [it]', Gustav wrote. 'From my entire recollections I cannot conceive that Heisenberg would have produced anti-Jewish sneers and obscenities and would have spat on the floor in front of his former professor, whom to the best of my knowledge he revered.'

For Rose this merely showed that Born was so embarrassed and upset that he hadn't even mentioned it to his son. Rose's view of Heisenberg is so negative that he finds the story entirely consistent with what he believes about Heisenberg's character. Among historians he is more or less alone in that belief.

that one cannot dismiss all that the scientists and other citizens tolerated as naïvety or patriotism. Born did share something of Heisenberg's
attitude that the barbarous excesses of the Nazis would soften with
time.

Paul Rosbaud, an editor of the KWG's scientific journal *Naturwissenschaften*, who had excellent contacts with many leading scientists
in Germany, was dismayed and rather disgusted by the lack of backbone he saw among German academics. As he later wrote,

> I remember one distinguished member of Göttingen University saying
> to me: 'If they should venture to break our university to pieces by
> expelling men such as James Franck, Born, Courant, Landau [the latter
> two mathematicians], we shall rise like one man to protest against it.'
> The next day, the newspapers reported that the same scientists and
> many others had been dismissed owing to their Jewish race and their
> disgraceful influence on universities and students. And all the other
> members of Göttingen University remained sitting and had forgotten
> their intention to rise and protest.

Rosbaud saw in this response an insidious blend of apathy, cravenness,
self-justification and latent anti-Semitism:

> The general excuse was: 'We could not dare to protest, though the
> expulsion of our Jewish colleagues is completely against our views and
> even against our conscience. We could not think of ourselves but of
> the higher purpose, the university, the academy. We had to avoid the
> possibility of these institutions having any trouble or their being closed.
> This was our first duty and so our personal views and interests, as well
> as those of our Jewish colleagues had to be kept in the background.'

The sense of helpless fatalism among the academic scientists seems
not so much misjudged as calculatedly self-serving. The Hungarian
physicist Leo Szilard, working at the University of Berlin in 1933 but
shortly to leave for England, expressed the situation very well:

> I noticed that the Germans always took a utilitarian point of view.
> They asked, 'Well, suppose I would oppose this, what good would I
> do? I wouldn't do very much good, I would just lose my influence.

Then why should I oppose it?' You see, the moral point of view was completely absent, or very weak, and every consideration was simply, what would be the predictable consequence of my action. And on that basis did I reach the conclusion in 1931 that Hitler would get into power, not because the forces of the Nazi revolution were so strong, but rather because I thought that there would be no resistance whatsoever.

But Rosbaud saw darker elements at work in the reactions of the academics:

Many of them added – and this was probably the first token of the beginning infection and confusion of mind – 'Besides, didn't they [the Jews] really go too far with their abstraction in science, and didn't they really go too far in accumulating Jewish collaborators? It is their own fault, and they must now pay for it. Perhaps they were really dangerous to our scientific life.'

Here he perceptively invokes a phenomenon now well attested: victims of discrimination become in fact despised, the object of anger and recrimination, considered to have 'brought it upon themselves'. Certainly, we should not complacently imagine that in all this there was anything uniquely Germanic. 'Would the populations of other countries have responded in more "honourable" fashion in similar circumstances?' asks Ian Kershaw. 'I suspect not.'

Meeting Hitler

It was customary for the president of the Kaiser Wilhelm Society to meet with any new head of state. So it was that on 16 May 1933 Max Planck went to see Adolf Hitler in Berlin.

Understanding the behaviour of the German physicists under the Nazis is rarely a matter of simply collating the documentary evidence and totting up episodes of compliance or resistance. Much of the story lies beneath the surface, in what is unspoken, in ambiguous phrases and subtle interpretations of apparently bland formalities and formulas, in the evasions and contradictions and accusations, the inability of the protagonists to articulate their emotions and motivations even in their own private correspondence. One finds oneself negotiating a coded language, seeking hints and clues about the true meaning. Key events and turning points become *Rashomon*-like narratives in which multiple viewpoints leave one despairing at ever deducing who did or said what, and why. One can, in consequence, tell pretty much whatever story one chooses, and people have done so. With this collection of conflicting accounts, one often has no option but to fall back on subjective assessment, seeking for consistencies and contradictions of character.

Planck's meeting with Hitler is one of these multiple narratives. No one knows exactly what transpired between the two men. There are even disagreements about how the encounter came about: was it an obligatory formality, or did Planck engineer the meeting in order to pursue some particular agenda? And if so, what was it? Was his main objective to appeal against the dismissals of the Jews? In any case, he did so.

Some accounts suggest that, confronted with Planck's entreaties, Hitler flew into a rage and the physicist meekly fled. Einstein even claimed he heard that Hitler had threatened the ageing scientist with the concentration camp. Other reports imply that the meeting was cordial throughout, and that Planck emerged satisfied that he had bought the (relative) autonomy and security of German physics at the cost of a servile, self-imposed *Gleichschaltung* – the alignment with Nazi doctrine that the party demanded in all aspects of German society.

Let's start with what Planck himself said about the event. In May 1947, just a few weeks before his death, he published in the *Physikalische Blätter*, the journal of the German Physical Society (Deutsche Physikalische Gesellschaft, DPG), an article called 'My visit to Hitler'. 'After Hitler came to power', he wrote,

I had the task as president of the KWG to wait upon the Führer. I thought to use the opportunity to say a word in favour of my Jewish colleague Fritz Haber, without whose process for making ammonia from atmospheric nitrogen the First World War would have been lost from the beginning. Hitler answered me with these words: 'I have nothing against the Jews themselves. But Jews are all Communists, and they are my enemies, against them I wage war.'

But surely there are, Planck alleged, 'all kinds of Jews, some valuable for mankind and others worthless, and among the first old families with the best German culture, and that distinction must be made'. To which Hitler responded: 'That is not right. A Jew is a Jew. All Jews stick together like leeches.'

But, Planck continued, 'it would be self-mutilation to make valuable Jews emigrate, since we need their scientific work'. Hitler said nothing in direct response to this,* but merely 'uttered some commonplaces' before falling into an unsettling disposition. 'People say that I suffer from a weakness of nerves', he advised Planck. 'That is slander. I have nerves of steel.' Whereupon the Führer slapped himself on the knee, spoke increasingly fast, 'and whipped himself into such a frenzy that I had no choice except to fall silent and leave'.

Many questions arise. How did Planck really feel about the Jews? Was Hitler truly so heedless of the damage his racial policies wreaked on science? And was Planck really left with no choice but to 'fall silent and leave' – to conclude that the best he could do was to administer the new laws as gently as possible? What else could one realistically expect of Planck and his colleagues in these circumstances?

It is important to recognize how Planck's article came about. It was solicited by the editor of the *Physikalische Blätter*, Ernst Brüche, as part of an effort to explain – and in part to exculpate – the actions of the German physicists before and during the war. Planck was very frail by this time, and the article was written with his help by his wife Marga, who edited the text so as to shield her husband from possible criticism.

What should we make of Planck's remark that some Jews are 'valuable for mankind and others worthless'? Was Planck suggesting that some Jews *specifically* are 'worthless for mankind', or implying that this is the case for any subset of humanity, including Jews? If in any event he allowed that some Jews had no value, did this reflect his personal view or was it just an attempt to pacify the Führer and obtain concessions? The question is complicated by the fact that Planck himself apparently did not originally phrase the issue in quite these terms. Instead of 'valuable' and 'worthless', he drew the distinction between 'Western' and 'Eastern' Jews, a standard formula at that time

---

* Other accounts make Hitler's response more melodramatically ominous: if Germany must do without science for a while, he is said to have intoned, so be it.

for distinguishing between assimilated and unassimilated Jews: those who had fitted into German society and those who remained 'alien'. In any event, this is a recurrent problem: how to assess statements of prejudice in the comments of non-Nazis forced to do business with their leaders. Is this merely a compromise (and if so, is it justifiable), or does it imply some acceptance of the ideas in both parties?

One of the oddest aspects of Planck's account – Hitler's claim to have 'nothing against the Jews' – is corroborated by Lotte Warburg, sister of the biochemist Otto, in her description of a visit from Erwin Schrödinger's wife Anny in July 1933. Frau Schrödinger told her host that Hitler

> said to Planck that he was not an anti-Semite, as people always label him; he is only against Communism, but the Jews have all become Communists. That is the only reason to fight them. Planck had the impression that Hitler is now very tired of the entire Jewish business, but that he cannot stop it.

We must of course be very wary of placing any interpretation on remarks framed by a psychotic mind. It's conceivable, however, that Hitler might have been artfully foreclosing the discussion. If Planck had come all ready to appeal against anti-Semitic discrimination, where could he go with that if Hitler proclaimed that he had nothing against the Jews after all?

Was that Planck's main intention for the meeting in any case? As we saw, Heisenberg wrote to Born two weeks afterwards to say that he understood Hitler to have promised Planck that beyond the Civil Service Laws, nothing else would be done that might hurt German science. This leads historian Helmuth Albrecht to conclude that Planck was essentially brokering a deal: if we go along with these laws, you'll leave us be. That interpretation, Albrecht says, is certainly consistent with the fact that state funding of the KWG increased subsequently. And Planck's later action suggests that he felt some arrangement had been reached, even if it was not the one he'd have privately preferred: he wrote to Hitler acceding euphemistically that the KWG was ready for 'the consolidation of available forces for an active contribution to construct our fatherland'. Or as the New York Times saw it in May 1933: 'German scientists rally behind Hitler.'

But Planck may not have really felt he had achieved anything worthwhile. The Jewish novelist Jakob Wassermann told his friend Thomas Mann in Switzerland that Planck had been 'utterly crushed' by the meeting, and that it had revealed to him the crude demagoguery of the new rulers: as Wassermann put it, 'disciplined thought must attend to the arrogant, dogmatic expectorations of a revolting dilettantism, bow, and withdraw'.

Nonetheless, Albrecht's view of the meeting between Planck and Hitler as an accommodation that bought some autonomy, allowing science to remain 'apolitical', seems to have been a perception that others besides Heisenberg shared. Obviously, such an outcome was sheer fantasy. It was worse than a fantasy: it was a deliberate delusion and a recipe for inaction. The Nazi state had no intention of granting scientists any exemption from *Gleichschaltung*. Perhaps their only respite was that, as Paul Rosbaud put it, 'Nobody of the Nazi leaders had any idea for what science can be used.' All the same, they treated the scientists like other academics, which meant insisting on empty, childish displays of loyalty, making students march in quasi-military parades, and when the time came, sending them all to the front to fight. Where science seemed able to serve the leaders' bidding, for example in chemistry (making armaments) or anthropology (devising crude, anti-Semitic racial doctrines), it did so. Experimental and classical physics was of value for aeronautics, ballistics and the creation of military instruments and weaponry. But until almost the eve of war, the new quantum, relativistic and nuclear physics did not appear to be of much use to anybody.

5 'Service to science must be service to the nation'

One aspect of Planck's account of his meeting with Hitler that we can accept without question is his concern about the fate of Fritz Haber, the director of the Kaiser Wilhelm Institute of Physical Chemistry and Electrochemistry in Dahlem. Despite having Jewish parents, Haber had been baptized and occasionally attended church. It made no difference within the dogma of Aryanism. The Jew-haters knew that their enemies never changed under the skin.

Haber had been immensely useful to imperial Germany. Not only, as Planck told Hitler, had he invented the process by which nitrogen was converted into ammonia, an essential precursor for explosives, but also he had masterminded the production of chlorine gas for chemical warfare.* No one could accuse Haber of lacking dedication to the military applications of chemistry. Shortly after chlorine was released on the battlefield at Ypres in 1915, he departed to supervise its use on the Eastern Front – the day after his first wife Clara (also a chemist) committed suicide by shooting herself with her husband's military pistol, apparently in shame and horror at the direction his research had taken. Yet Haber expressed no regrets about this wartime work, and poison-gas research continued at his Berlin institute after the war. The cyanide gas Zyklon A was developed there in the 1920s as an insecticide; the Nazis found another application for the later modification, Zyklon B.

Haber's work on chemical weapons has often been presented as evidence that he was an amoral monster – and as such, an aberration

---

* Somewhat less useful to his homeland were Haber's failed attempts in the 1920s to develop a method for extracting gold from seawater to help pay off Germany's war reparations.

in science. But it was an almost universally accepted duty to make one's services available to the military during wartime. Haber's war work gained him the respect of his contemporaries – it made him a noble German, and no one doubted his patriotism. Besides, he hoped that the shock of chemical warfare would bring an end to the stalemate of trench warfare, forcing an early resolution to the war and so ultimately saving lives.

In 1933 many of the Kaiser Wilhelm Institutes were in an ambiguous position with regard to the Civil Service Law. Since they were funded by a partnership of government and industry, most of their staff were not exactly government employees. Haber's KWIPC was different, however, for it was under direct government control. The new law did not threaten Haber personally since his war record exempted him, but he was told to purge his staff of Jews. The institute had a high proportion of 'non-Aryan' researchers – about one in four, which to anti-Semites was more evidence of how the Jews looked after their own. It's understandable that they should do so, averred Bernhard Rust, head of the scientific branch of the Reich Education Ministry (REM), who was responsible for implementing the changes. But, he insisted, 'I cannot allow it . . . We must have a new Aryan generation at the universities, or else we will lose the future.' Rust assured the good German Jews that 'I deeply feel the tragedy of persons who inwardly want to consider themselves part of the German *Volk* community and work within it . . . But the principle must be carried out for the sake of the future.'

Ordered to dismiss many of his key staff members, Haber felt that the only honourable course was to tender his own resignation too. He stepped down with great dignity, writing to Rust in April:

> My tradition requires of me that in my scientific position I consider only the professional accomplishments and character of applicants when I choose my co-workers, without asking about their racial make-up. You cannot expect that a sixty-five-year-old man will change this way of thinking which has guided him for thirty-nine years of university life.

The Nazis 'stand before the pieces of a broken glass', the demoralized Haber told the Rockefeller's Warren Weaver in May. 'They now realize that they did not really wish to break it – and they don't know what to do with the fragments.' Weaver saw Haber as 'a pathetic yet noble figure. He has saved out of the wreck the only thing he could possibly save – his own self-respect.'

Remembrance day

Haber's resignation left Planck distraught, but his response shows how a slavish devotion to an obsolete sense of propriety paralysed him. 'What should I do?' he asked Lise Meitner when she protested the injustice. 'It is the law.' Planck knew that the legality of the dismissals did not make them right – but in his view it made them incontestable. As Alan Beyerchen puts it, 'One was faced with the contradictory position of protesting the illegality of the law, a concept which might make sense in Anglo-Saxon countries but did not in Germany.'

The fiction of the KWG's autonomy was brought home to Planck when it came to selecting Haber's successor. Planck proposed to Rust that this be Otto Hahn. Instead Rust appointed August Gerhart Jander, a rather undistinguished chemistry professor at Göttingen but, crucially, a loyal party member. He was to be assisted – in effect, commanded – by Rust's deputy Rudolf Mentzel, who thereafter turned up at meetings of the KWG senate in his SS uniform. Jander was ineffectual, but in 1935 he was replaced by Peter Adolf Thiessen, an 'Old Fighter' of the Nazi Party and a wholly competent scientist, who turned the institute into an efficient instrument of the regime. Its work became increasingly focused on chemical warfare, while evening gatherings and camps for 'the deepening of comradeship' were toasted with tankards of foaming beer.

Wasn't this an indication of what would befall all of German science if its leading representatives resigned their posts – that it would either be run by incompetents or become subservient to the Nazi agenda? That is what Planck and Heisenberg feared. For Heisenberg, to simply down tools and leave the country was a dereliction of duty, not a moral act of protest.

Poor Haber left Germany a broken man. The pain of rejection by the country he loved is clear in the plaintive words he wrote to Carl Bosch from England in December 1933: 'I never did anything, never even said a single word, that could warrant making me an enemy of those now ruling Germany.' The following month he died in Switzerland of heart disease. The KWG's *Naturwissenschaften*, which staunchly resisted *Gleichschaltung*, carried an obituary from Laue in which he insisted on Haber's place in German culture: 'He was one of our own', he wrote.

Planck, Laue and others decided there should be a memorial meeting on the first anniversary of Haber's death, 29 January 1935, to be held at the KWG's headquarters at Harnack House in Berlin. But researchers at most of the Kaiser Wilhelm Institutes were not formally included in the official ban, and several of them came, knowing that this would be duly reported to the authorities. They included Carl Bosch, Lise Meitner, Otto Hahn, their students Fritz Strassmann and Max Delbrück, and Planck himself. No one was quite sure if the gathering would be prevented by force, but in the event it was well attended and passed peacefully. Planck said a few words of appreciation to their former colleague: 'Haber remained true to us', he proclaimed, 'we will remain true to him.'

The Haber memorial has sometimes been paraded as evidence that German scientists did mount opposition to the Nazis. But it wasn't really that at all, less still a symbolic protest against anti-Semitism. For Planck it was simply a proper observance of tradition: in requesting permission for the event from Bernhard Rust at the REM, he defended it as an 'old custom' with no political connotations. Although Rust replied sternly, saying 'Haber has done a lot for science and for Germany, but the NSDAP has done a lot more', he gave Planck permission to proceed with the gathering. And once the ministry forbade any university professors from attending, saying that this would be regarded as a provocative act, the academics stayed away.* Even Laue complied, rightly assuming that Nazi spies would be at the event.

So the Haber memorial was, in effect, state-sanctioned, albeit reluctantly and accompanied by a predictable refusal of permission to publish the proceedings. Historian Joseph Haberer delivers a damning but at least partly warranted judgement in calling the Haber memorial 'a device for justifying the collapse of civic courage'. The gathering showed once more how the National Socialists might tolerate

---

* This action was taken by one of Rust's subordinates, without his knowledge, while he was ill.

what they regarded as professional antics and rituals among the scientists, perhaps recognizing that by discharging their grievances in an apolitical manner these insignificant concessions could facilitate a more general compliance.

And Planck gave ample evidence of his willingness to compromise. He spoke again of Haber's accomplishments on the twenty-fifth anniversary of the KWG in 1936 (and was reprimanded for it), but he kept Haber's name mostly absent from the published record of that event. Planck also marked this celebration with a telegram to Hitler thanking him for his 'benevolent protection of German science'.

Under National Socialism, Planck later claimed, the KWG found it expedient to behave 'like a tree in the wind', bending when necessary but becoming upright again when the pressure passed. He never really saw that the bending was all the Nazis cared about.

The end of mathematics

Haber was not the only Jew to resign in protest at the Civil Service Law despite having no formal obligation to do so. The German scientific community was equally shocked by the decision of James Franck to vacate his chair as professor of physics at Göttingen. Franck, who had won the Nobel Prize in 1925 for his research on the quantum theory of the atom, had been awarded two Iron Crosses in the First World War, and so was an exempt veteran by any standards. But he explained that he could not remain employed by a state that made his children second-class citizens, and neither would he stand back and watch others being unjustly dismissed. Some of his colleagues tried to dissuade Franck on the grounds that, as the young physicist Rudolf Hilsch put it, 'nothing is eaten as hot as it is cooked': the heat would pass. But a greater number of academics deplored this blatantly 'political' act, especially when Franck's resignation letter was published in the *Göttinger Zeitung*. Forty-two members of the Göttingen staff signed a petition calling it 'equivalent to an act of sabotage'.

Franck didn't leave the country at once: he stayed in Göttingen, hoping to find a non-academic post. 'There is, of course, no chance of this', Max Born wrote to Einstein in June. By this time Born himself had already decided to quit Göttingen and was in the Italian Tyrol considering various job offers in the United States and France. Shy of publicity, he did not wish to pursue Franck's bold course. 'I would not have the nerve to do it', he confessed to Einstein, 'nor can I see the point of it.' 'As regards my wife and children', he added,

> they have only become conscious of being Jews or 'non-Aryans' (to use the delightful technical term) during the last few months, and I myself have never felt particularly Jewish. Now, of course, I am extremely conscious of it, not only because we are considered to be so, but because oppression and injustice provoke me to anger and resistance.

Also ousted at Göttingen was the mathematician Richard Courant. But he did not go easily, electing instead to contest the situation. It was hopeless, all the more so after a smear campaign alleged he had been a Communist. What eventually sealed Courant's decision to emigrate, however, was his fears for his family – not so much about their being in physical danger, but about the danger of infection from the poison seeping into German society. 'My youngest son', he later wrote, 'did not seem able to understand why he should not be in the Hitler Youth, too.'

Göttingen had been a jewel of mathematical physics, but the dismissals and resignations all but destroyed its scientific standing. Others who left included the Hungarian physical chemist Edward Teller, the mathematician Hermann Weyl, Franck's son-in-law Arthur von Hippel, the naturalized Russian Jew Eugene Rabinowitch, and the physicist Heinrich Kuhn. Many of these, like Franck himself (who settled at the University of Chicago) would carry out vital war research for the Allies, particularly on the Manhattan Project. Teller would become one of the key agitators for the post-war development of the thermonuclear hydrogen bomb. Shortly after the exodus, the mathematician David Hilbert found himself seated next to Bernhard Rust at a banquet. The minister asked him, 'And how is mathematics in Göttingen now that it has been freed of the Jewish influence?' Hilbert replied, 'Mathematics in Göttingen? There is really none any more.'

We have seen already how 'non-Jews' mostly accepted these events without demur – some through fear for their position and prospects, some from fatalism or a determination to 'preserve German science', some because they stood to benefit from or simply agreed with the new law. Others offered lame excuses. Warren Weaver wrote of the KWG's secretary Friedrich Glum that he 'makes his defense of the situation with his eyes down on the table. His defense is moreover unimpressive and shallow.' In response to Weaver's protestations, Glum retorted by citing American prejudice against black people. Weaver pointed out that the difference was that he and other liberals did not endorse, defend or excuse this. Glum fell silent. 'Only in the case of a few really noble and courageous men, such as Planck, does one meet sincerity or anything approaching frankness', wrote Weaver.

Even the scientists who 'assisted' or 'defended' their Jewish colleagues failed to see that their actions ultimately facilitated and even endorsed the process. They would, with expressions of regret, help Jewish scientists find positions abroad, but would offer little support for those rare individuals like Courant who tried to stay. By participating in the process of finding replacements, they were tacitly accepting the legitimacy of the reasons for the vacancy. Heisenberg, having failed to persuade Born to stay in Göttingen, accepted an appointment to his vacated post, although political machinations ultimately prevented him from taking it.

Lacking any experience of organized resistance to the state, the scientists had no idea what else they could do. They hoped that their compliance would limit state intrusion. 'The watchword', according to Beyerchen,

> was that those who could should stay. The goals of these leaders were to minimize individual hardships, reverse the dismissals and resignations when possible and, above all, to maintain the international standing of German science . . . The worst of National Socialism would pass, these men felt, but the importance of science for Germany's reputation would endure.

They discovered soon enough that the worst was not going to pass quickly. After August 1934 all civil servants were required to sign an oath of allegiance to Hitler in person: the final seal on the Führer state. Heisenberg and Debye in Leipzig signed in January 1935, their university having been Nazified with barely a murmur. 'Swastikas can be seen everywhere', wrote the visiting Italian scientist Ettore Majorana, adding that 'the Jewish persecutions delight most Aryans'.

Students began to abuse remaining Jewish professors openly and mounted demonstrations against the 'non-German spirit'. The university was given a new constitution which adopted the 'Führer principle' (incontestable leadership by a single individual), and a new rector, Arthur Golf, who stated that students and professors would thenceforth be 'comrades under Hitler'. The 'Commitment of the Professors at German Universities and Colleges to Adolf Hitler' had been celebrated in November 1933 at a meeting of the National Socialist Teachers League in Leipzig's Albert Halle, organized by Johannes Stark, with an address by the rector of Freiburg, philosopher Martin Heidegger.

By 1935, one in five scientists in Germany – one in four in the case of physicists – had been dismissed. And several (but by no means most) positions of power and influence had been filled by mediocre individuals promoted because of their obedience to the party. Moreover, the Nazis seemed to begin insisting not just on who did science, but on what science was done. In June 1933 Interior Minister Frick had proclaimed that 'With all respect for the freedom of science, let us postulate that service to science must be service to the nation and that scientific achievements are worthless when they cannot be utilized for the culture of the people.' In a speech to professors in Munich, the Bavarian state minister for instruction and culture advised that 'From now on, the question for you is not to determine whether something is true, but to determine whether it is in the spirit of the National Socialist revolution.'

If that sounds like anathema to good science, however, in practice such empty slogans made little difference: the Nazi leaders were in no position to evaluate these distinctions, and weren't greatly interested in them anyway. As we shall see in Chapter 6, the encroachment of Nazi ideology in physics was not a state-sanctioned enterprise but an ultimately fruitless attempt at self-promotion by a few eminent yet embittered individuals.

Einstein expunged

When Hitler was made chancellor in January 1933, Albert Einstein was visiting the California Institute of Technology. On 10 March he announced that he would not return to live in his native country: 'As long as I have any choice in the matter, I shall live only in a country where civil liberty, tolerance, and equality before the law prevail . . . These conditions do not exist in Germany at the present time.' He returned briefly to Europe before settling at the Institute for Advanced Study in Princeton, and in May he wrote from Oxford to Max Born in the Italian Tyrol. What he said would surely have scandalized his patriotic colleagues:

> You know, I think, that I have never had a particularly favourable opinion of the Germans (morally and politically speaking). But I must confess that the degree of their brutality and cowardice came as something of a surprise to me.

He was talking not just about the Nazis but also his former friends and associates, who had decided that Einstein's refusal to come back to a country that excluded Jews from full citizenship was an act of gross treason. The Prussian Academy of Sciences was outraged by that. As its president, Max Planck was expected to write a letter of condemnation to his friend. He did so, his feeble premise being that Einstein's comments and actions were not *useful*:

> By your efforts, your racial and religious brethren will not get relief from their situation, which is already difficult enough, but rather they will be pressed the more.

In other words, Planck was insisting that Einstein should accept anti-Semitic discrimination in silence so that it did not get even worse. After all, Planck had once told Einstein, the value of an act lies with its consequences and not its motives. While that might be debatable philosophically and theologically, it was a useful position for the German physicists, who could then justify any decision on the grounds that it alone held out hope of making things better, or at least not making them worse. Acting on principle was, in this view, egotistical and irresponsible. Even Laue concurred, writing to Einstein that 'Here they are making nearly the entirety of German academics responsible when you do something political.'

But Einstein would not back down. 'I do not share your view', he told Laue, that the scientist should observe silence in political matters, i.e. human affairs in the broader sense . . . Does not such restraint signify a lack of responsibility?. . . I do not regret one word of what I have said and am of the belief that my actions have served mankind.

This is really the point: for Einstein, 'political' meant 'human affairs in the broader sense', and thus, questions of right or wrong, fair or unjust, kind or cruel. Laue, as we have seen already, did not turn a blind eye to these matters – he found the Nazis repugnant and challenged them on many occasions. He did not lack responsibility. But he would not have considered these brave acts of defiance to be 'political'. Even he could not see beyond the narrow conception to which the German academics clung, in which 'political' meant something close to 'unpatriotic' – a word or action that did not simply defy some ugly or stupid decision by a government official but which questioned the legitimacy of the German state. Laue was unable to connect his strong sense of personal morality to a duty to 'serve mankind'. The Fatherland had laid claim to any such duty. For Planck that claim was incapacitating.

Besides, while Planck and Laue may not have been right that Einstein was aggravating the situation, they had reason to think so. Consider what Joseph Goebbels had to say on the matter: 'The Jews in Germany can thank refugees like Einstein for the fact that they themselves are today – completely legitimately and legally – being called to account.' It was that 'legitimately and legally' that stymied Planck.*

When Einstein failed to explain his conduct to the Prussian Academy, the presiding secretary, meteorologist Heinrich von Ficker, urged Planck to demand Einstein's resignation. He did this too – but

---

* Einstein was not blind to this possibility. In June 1933 he wrote to Laue: 'I have learned that my unclear relationship to those German organizations which still include my name in the list of members could cause problems for my friends in Germany. For this reason, I would like to ask you to make sure that my name is removed from the lists of these organizations. These include, for example, the German Physical Society [DPG] . . . I am explicitly empowering you to do this for me.' Laue's response was surely heartfelt: 'Although I am very thankful that you are trying to make things as easy as possible for us, I nevertheless could not do these . . . things without the deepest sadness.' The DPG, however, accepted the resignation without comment, as though it was nothing exceptional. Einstein never rejoined after the war.

Einstein got there first, tendering his resignation before Planck's sheepish letter arrived. That angered the Prussian minister for cultural affairs – none other than Bernhard Rust, who had not yet been appointed to the REM. Rust demanded that the academy discipline Einstein for his 'agitation'. Hadn't he been saying vile things about Germany? Admittedly, the only evidence of this came from American newspaper reports, but nonetheless . . . Another of the academy's secretaries, the orientalist Ernst Heymann, drafted a hurried statement while Planck was away on holiday in Sicily:

> The Prussian Academy of Sciences heard with indignation from the newspapers of Albert Einstein's participation in the atrocity-mongering in France and America . . . The Prussian Academy of Sciences is particularly distressed by Einstein's activities as an agitator in foreign countries, as it and its members have always felt themselves bound by the closest ties to the Prussian state and, while abstaining strictly from all political partisanship, have always stressed and remained faithful to the national idea. It has therefore no reason to regret Einstein's withdrawal.

Heymann saw no irony in issuing this declaration of political impartiality on 1 April, the day on which the Führer called for a boycott of Jewish businesses. Indeed, the REM had hoped to parade Einstein's dismissal as a trophy in this anti-Semitic spree.

Laue was affronted by the slur to Einstein's reputation, and demanded a meeting of the academy's committee to review the matter. But the committee merely endorsed the wording, in what was for Laue 'one of the most appalling experiences of my life'. Planck, however, was aware that posterity might judge Einstein's withdrawal very differently, and in the minutes of the academy's meeting on 11 April he noted the unquestioned and abiding importance of Einstein's work before implicitly reversing Heymann's judgement: 'Therefore it is . . . deeply to be regretted that Einstein has by his own political behaviour made his continuation in the academy impossible.' In his letter to Rust informing him of Einstein's resignation, Planck wrote 'I am convinced that in the future the name of Einstein will be honoured as one of the most brilliant intellects that has ever shone in our academy.' Such attempts to claw back credibility doubtless undermined whatever political credit Planck had gained from Einstein's resignation in the first place.

On 5 April Einstein responded forcefully to Heymann's accusations with a public statement:

> I hereby declare that I have never taken any part in atrocity-mongering, and I must add that I have seen nothing of any such mongering elsewhere. In general, people have contented themselves with reproducing and commenting on the official statements and orders of responsible members of the German government, together with the program of the annihilation of the German Jews by economic methods. The statements I have issued to the press were concerned with my intention to resign my position in the academy and renounce my Prussian citizenship; I gave as my reason for these steps that I did not wish to live in a country where the individual does not enjoy equality before the law, and freedom of speech and teaching. Further, I described the present state of affairs in Germany as a state of psychic distemper in the masses and made some remarks about its causes . . . I am ready to stand by every word I have published.

The following day Einstein also wrote privately to Planck, making the same defence but in milder terms. He tried to make his friend see the matter for what it was by removing the Jewish context, as if trying to help a child to put himself in another's shoes:

> I ask you to imagine yourself for the moment in this situation: assume that you were a university professor in Prague and that a government came into power which would deprive Czechs of German origin of their livelihood and at the same time employ crude methods to prevent them from leaving the country . . . Would you then deem it decent to remain a silent witness to such developments without raising your voice in support of those who are being persecuted? And is not the destruction of the German Jews by starvation the official program of the present German government?

Einstein was no saint, yet there is immense forbearance in his refusal to hold Planck's actions against him: 'I am happy that you have nevertheless approached me as an old friend and that, in spite of severe pressures from without, the relationship between us has not been affected. It remains as fine and genuine as ever.'

The Prussian Academy, meanwhile, was unrelenting. Ficker doubtless knew he had no case to assert, but he responded to Einstein's declaration with moustache-twirling bluster:

> We had confidently expected that one who had belonged to our academy for so long would have ranged himself, irrespective of his own political sympathies, on the side of the defenders of our nation against the flood of lies which has been let loose upon it . . . Instead of which your testimony has served as a handle to the enemies not merely of the present government but of the German people.

In other words, patriotism should override any 'political sympathies' on behalf of oppressed Jews. Ficker and Heymann added in a separate letter that even if Einstein hadn't been involved in 'atrocity-mongering', he should at least have done something 'to counteract unjust suspicions and slanders'. Ficker does not appear to have been a Nazi sympathizer, neither is he here attempting to cover up or justify anti-Semitism. Rather, for him – and, one must conclude, for Planck – this is beside the point. He is insisting that, however one feels about the political 'Jewish question', the first duty is to defend Germany's honour.

To behave in that way, Einstein told Ficker and Heymann by return of post, 'would have been equivalent to a repudiation of all those notions of justice and liberty for which I have stood all my life . . . By giving such testimony in the present circumstances I should have been contributing . . . to moral corruption and the destruction of all existing cultural values.' That, he added, was why he resigned, 'and your letter only shows me how right I was to do so'.

As a result of Einstein's resignation, the Bavarian Academy of Sciences, of which Einstein was also a member, took fright and sent him a nervous letter from which one can sense the academicians fretting that they might be deemed politically unsound if they do not follow the Prussians' lead.* They asked Einstein to clarify 'how you

---

* Such gestures did nothing to preserve the Bavarian Academy's independence, for in 1936 the Reich Education Ministry stipulated that it would thereafter appoint the president and secretaries directly.

envisage your relations with our academy after what has passed between yourself and the Prussian Academy'. Einstein's response might be paraphrased as 'now you mention it, I don't particularly want to be a part of your organization either'. But this, he said, is for a different reason:

> The primary duty of an academy is to further and protect the scientific life of a country. And yet the learned societies of Germany have, to the best of my knowledge, stood by and said nothing while a not inconsiderable proportion of German scholars and students and also of academically trained professionals have been deprived of all chance of getting employment or earning a living in Germany. I do not wish to belong to any society which behaved in such a manner, even if it does so under external pressure.

These remarks reveal that it is not just in retrospect, within a different social and political context, that it becomes possible to speak of the morality of the situation. Einstein is not demanding that anyone resign or even refuse outright to comply with the Nazi strictures, but merely that one should not carry on as though all is well. Planck's view that nothing *could* be done, and the attitude of Ficker and Heymann that nothing *needed* to be done, were in the end indistinguishable in their consequences. 'When faced with a choice between endangering their academy or acquiescing in the racist purge of the Prussian Academy of Sciences', writes Mark Walker, 'the academy scientists surrendered their independence and became accomplices by helping the National Socialist state force the Jewish scientists out of the academy. No "Aryan" scientists resigned in protest. Indeed there is no record of a scientist even considering resignation.' This statement probably reflects Planck's position, but is perhaps too generous overall: there seems good reason to suppose that many others in the academy did not just acquiesce in the anti-Semitic purge but actively approved of it.

Politically worthless

Planck's reluctant compliance did him no favours with anyone. It was clear that his heart wasn't in the persecution of Einstein, all the more so when he proposed Laue, Einstein's friend and a committed supporter of his theories, as Einstein's successor for the non-teaching professorship that the academy awarded. Philipp Lenard, a fierce critic of Einstein both scientifically and politically, objected, calling Planck 'politically so worthless a character', while Lenard's associate Johannes Stark commented that 'if Planck and Laue retain influence, it will have a worse effect than if Einstein himself were there'.

Stark decided that to counteract this pernicious influence he would seek government support to get himself elected to the academy. The proposal for Stark's membership was drawn up by the experimental physicist Friedrich Paschen, who had previously joined Laue in *opposing* Heymann's cavalier press statement about Einstein. Paschen warned Ficker that to object to Stark's application would be 'tactically a false step and even dangerous'. Laue nonetheless succeeded in getting the motion blocked in December. When as a result Stark spitefully fired Laue from his consultancy to the Physical and Technical Institute of the German Reich (PTR) in Berlin, of which Stark was the president, the previously submissive Ficker circulated some of the vicious things that Stark had been saying about Planck, Laue, Haber and others. In January 1934 his nomination was withdrawn. The episode makes clear how hard it is to discern where the battle lines were drawn – critics of Einstein were not necessarily friends of outright Nazis like Stark, while Einstein's sometime supporters were not immune to political expediency.

The exclusion of Stark was a pyrrhic victory for those who wished to prevent the Nazification of the Prussian Academy. The National Socialists encouraged academicians to apply for party membership by offering the bait of preferential consideration for the best Civil Service jobs. From 1934 the academy signed off its correspondence with 'Heil Hitler!', as most civil servants were required to do. Servile gestures of loyalty, such as communal listening to Hitler's radio broadcasts, were demanded and observed. Planck unsuccessfully contested the election of party members in 1937, most notably the mathematician Theodor Vahlen, whose journal *Deutsche Mathematik* aimed to establish his subject as an 'Aryan' discipline purged of Jewish influence. After that, any semblance of autonomy vanished. In 1938 Rust removed the last vestiges of democratic process, replacing the academy's committee with governance by the Führer principle and giving his ministry the power to appoint or dismiss members as it saw fit. When Rust made

Vahlen president in 1939 (with Vahlen's fellow *Deutsche Mathematiker* Ludwig Bieberach as secretary), the game was over: the academy ceased to function as a serious scientific body and became an organ of the Nazi state. Planck's accommodations had been for nothing – indeed, eventually he gave up his opposition to the changes and announced that the academy should put their trust in Vahlen. At no stage did Planck appear to consider resigning his membership.*

The ejection of Einstein from the Prussian Academy of Sciences shows Planck at his most compromised. In some ways his actions look all the worse for the fact that Planck was *not* one of the ultra-nationalist anti-Semites calling for Einstein's exclusion: at some level he surely knew the injustice of it all. But his position is more tragic than despicable: he could not imagine what else to do. According to the Austrian physicist and philosopher Philipp Frank, who emigrated from Prague to the United States in 1938,

> Max Planck was one of the German professors who repeatedly asserted that the new rulers were pursuing a great and noble aim. We scientists who do not understand politics, ought not to make any difficulties for them. It is our task to see to it that as far as possible individual scientists suffer as few hardships as possible, and above all we should do everything in our power to maintain the high level of science in Germany. At least envious foreigners should not notice that a lowering of the level is taking place anywhere in our country.

This equivocation in a fundamentally decent and honest but inflexible man is evoked pathetically in a description by Paul Ewald of Planck's predicament the following year, symbolizing what must have been for him a permanent inner struggle:

> I think it was on the occasion of the opening of the Kaiser Wilhelm Institute of Metals in Stuttgart, and Planck as president of the KWG came to the opening. And he had to give the talk, and this must have

* After the war, the Prussian Academy of Sciences exemplified the manner in which formerly Nazified state bodies accommodated themselves to the new regime with rather little fuss. It became the Berlin Academy of Sciences in 1945, then the German Academy of Sciences, and played an important role in the science of the German Democratic Republic until being dissolved shortly after reunification.

been in 1934, and we were all staring at Planck, waiting to see what he would do at the opening, because at that time it was prescribed officially that you had to open such addresses with 'Heil Hitler'. Well, Planck stood on the rostrum and lifted his hand half high, and let it sink again. He did it a second time. Then finally the hand came up, and he said 'Heil Hitler.'. . . Looking back, it was the only thing you could do if you didn't want to jeopardize the whole KWG.

Everyone in German science was 'staring at Planck, waiting to see what he would do'. And what he did was apparently to endorse the feeblest of responses, based on considerations of what was allegedly 'good for Germany' in the long term. He had no plan, no moral compass beyond an innate goodness of heart, no precedent or historical model – nothing to guide him through the catastrophe that had engulfed him, and which in the end destroyed him.

6 'There is very likely a Nordic science'

Anti-Semitism did not just deprive German physics of some of its most valuable researchers. It also threatened to prescribe what kind of physics one could and could not do. For Nazi ideology was not merely a question of who should be allowed to live and work freely in the German state – like a virus, it worked its way into the very fabric of intellectual life. Shortly after the boycott of Jewish businesses at the start of April 1933, the Nazified German Students Association declared that literature should be cleansed of the 'un-German spirit', resulting on 10 May in the ritualistic burning of tens of thousands of books marred by Jewish intellectualism. These included works by Sigmund Freud, Bertolt Brecht, Karl Marx, Stefan Zweig and Walter Benjamin: books full of corrupt, unthinkable ideas. Into some of these pyres, baying students threw the books of Albert Einstein.

It was one thing to say that art was decadent – that its elitist abstraction or lurid imagery would lead people astray. And the 'depraved' sexuality saturating the pages of Freud's works was self-evidently contaminating. But how could a scientific theory be objectionable? How could one even develop a pseudo-moralistic position on a notion that was objectively right or wrong? Besides, hadn't Einstein's relativity been proven? What did it even mean to say that science could be subverted by the 'Jewish spirit'?

It would be absurd, of course, to suppose that most of the book-burners had given these questions a moment's thought. The simple fact was that Einstein was a prominent Jew, and his thoughts therefore fit for the bonfire. But Einstein's theory *was* attacked on racial grounds. This assault came not by asinine ideologues in the party whose knowledge of science extended no further than a belief in fairy tales about 'cosmic ice', nor from individuals on the scientific fringe seeking official approval and support. It was orchestrated by two Nobel laureates in physics, who devised a full-blown thesis (it can't be dignified by calling it a theory) on how stereotypical racial features are exhibited in scientific thinking. They were Philipp Lenard and Johannes Stark, and they wanted to become the new Führers of German physics.

The story is ugly, sad, at times comic. It illustrates the complicated interactions between science and politics in Nazi Germany, for although one might expect the 'Aryan physics' (*Deutsche Physik*) of Stark and Lenard to have been welcomed by the National Socialists, its reception in official circles was decidedly mixed, and in the end it was ignored. The case of *Deutsche Physik* reveals how much of what went on in the Nazi state depended on how you played your cards rather than on what sort of hand you held. It shows how the German scientists' pretensions of being 'apolitical' did not prevent politics from infecting scientific ideas themselves, and almost overwhelming them. Perhaps most importantly, the story explodes the comforting myth that science offers insulation against profound irrationality and extremism.

Against relativity

Lenard's anti-Semitism festered for years before the Nazi era, and as was the case with many other haters of Jews his antipathy was fuelled by a sense of exclusion and injustice. The fact is that Lenard was a rather unremarkable man: an excellent experimental scientist in his heyday, but of limited intellectual depth, and emotionally and imaginatively stunted. When circumstances contrived to carry him further than his talents should have permitted, he was forced to attribute his shortcomings to the deceptions and foolishness of others. This combination of prestige and deluded self-image is invariably poisonous. There is no better example than Lenard to show that a Nobel Prize is no guarantee of wisdom, humanity or greatness of any sort, and that, strange as it may seem, the award can occasionally provoke feelings of inadequacy.

Lenard was given the prize in 1905 for his studies of cathode rays, the 'radiation' emitted from hot metals. They were manifested as a glow that emerged from a negatively charged metal plate (cathode) inside a sealed, evacuated 'cathode-ray tube' and made its way to a positively charged plate. Directed on to the glass walls of the tube – or as researchers discovered, on to sheets of particular minerals – the cathode rays stimulated bright fluorescence. Like his mentor Heinrich Hertz at the University of Bonn, Lenard at first believed these rays to be fluctuations in the ether – like light, as it was then conceptualized. But while J. J. Thomson, director of the Cavendish Laboratory in Cambridge, noted in 1897 that this was 'the almost unanimous opinion of German physicists', he had results that implied otherwise. Thomson showed that cathode rays have negative electric charge, being deflected by electric and magnetic fields, and he concluded that they were in fact streams of particles. They were given the name proposed some years earlier by the Irish physicist George Johnstone Stoney for the smallest possible unit of electrical charge: electrons. As Lenard put it, electrons are the 'quanta of electricity'.

Lenard discovered how to enable cathode rays to escape from the vacuum chamber in which they were created, so that they could be examined more closely. He also investigated the photoelectric effect – the expulsion of electrons from metals irradiated with ultraviolet light – and discovered that the energy of these electrons did not depend on the intensity of the light but only on its wavelength. When Einstein explained this result in 1905 in terms of Planck's quantum hypothesis (see page 13), Lenard felt that his discovery had been stolen. That bitterness deepened when Einstein was awarded the 1921 Nobel Prize in Physics for his work on the photoelectric effect.

This was not Lenard's only early source of resentment. He felt that he should have discovered X-rays before Wilhelm Röntgen (page 15), and was sure that he would have done so if the jealousies of senior professors had not denied him better opportunities. And hadn't he offered Röntgen advice about constructing the cathode-ray tube used for this discovery, which Röntgen didn't even have the good grace to acknowledge?

But if the German professors selfishly and unjustly hid their intellectual debts, the English were worse. Thomson should have given him more credit for his work on the photoelectric effect, for instance. This, however, was no more than one could expect from a nation of vulgar materialists – Lenard would surely have sympathy with Napoleon's remark about shopkeepers – who knew nothing of the heroic, selfless Germanic *Kultur*. James Franck later claimed that, when he was fighting at the front in the First World War, Lenard wrote to him expressing his hope that the defeat of the English would make amends for their never having cited him decently.

An operation for an illness of the lymph nodes around 1907 left Lenard less able to work, and contributed to his difficulties in keeping up with the latest developments in physics. Because he was not mathematically adept, he could not get to grips with relativity or quantum theory. As a result, he decided they were nonsense. The fact that this nonsense – whose premier architect was Einstein – was being accepted and acclaimed by physicists all over the world must therefore be the result of a conspiracy. And conspiracies and cabals were the speciality of Jews.

Einstein was the embodiment of all that Lenard detested. Where Lenard was a militaristic nationalist, Einstein was a pacifistic internationalist. Einstein was feted everywhere, while Lenard's great merits seemed to have been forgotten. Worse, Einstein was celebrated most of all in England! And he hawked a brand of theoretical physics that frankly baffled Lenard. How convenient, then, that Einstein was a Jew, so that all of these deplorable traits could be labelled Semitic. (Of course, many of Einstein's supporters were not Jewish, but as we shall see, Lenard and his ilk later contrived to make them 'honorary Jews'.) Lenard decided that relativity was a 'Jewish fraud' and that anything important in the theory had been discovered already by 'Aryans'.*

Lenard criticized the theory of relativity as early as 1910, but it was not until the 1920s that his attacks began to incorporate explicitly racial elements. He started to develop the notion that there was a Jewish way of doing science, which involved spinning webs of abstract theory that lacked any roots in the firm and fertile soil of experimental work. The Jews, he said, turn debates about objective questions into personal disputes. Ironically, this supposed preference of 'Aryans' for hale and hearty experiment went hand in hand with the kind of Romantic mysticism that infuses Nazi philosophy, such as it is. Lenard approved of the animistic *Naturphilosophie* of Goethe and Schelling,

---

* In particular, Lenard began the myth that the theory of relativity had been devised by the Austrian physicist Friedrich Hasenöhrl – a story still popular with Einstein's cranky detractors today.

the belief in a spirit that animated all of nature. This pervasive soul of nature was the wellspring of science itself – and only Aryans, said Lenard, understood this: 'It was precisely the yearning of Nordic man to investigate a hypothetical interconnectedness in nature which was the origin of natural science.'

Lenard persisted in believing in the light-bearing ether that Einstein had rejected, saying cryptically that this elusive medium 'seems already to indicate the limits of the comprehensible'. He lamented the encroachment of technology in modern life: an expression, he said, of the kind of materialism that infected both Communism and the Jewish spirit, the twin enemies of German greatness. Materialistic natural science had eclipsed the 'spiritual sciences', giving rise to the 'arrogant delusion' that humankind can achieve the 'mastery of nature'. 'That influence has been strengthened by the all-corrupting foreign spirit permeating physics and mathematics', he wrote – 'foreign' here meaning, of course, Jewish.

The enthusiasm of the Nazi regime for this brand of mysticism and pseudoscience has been well documented, although perhaps not enough has yet been made of the resonances between fascism, *Naturphilosophie*, the cultish mysticism of Rudolf Steiner* and anthroposophy, and the cosy certainties of some New Age beliefs. Reified worship of nature (as opposed to respect for it) has always teetered on the brink of a fundamentally fascist ideology. Several Nazi leaders, including Hitler and Himmler, endorsed the ridiculous 'cosmic ice' theory of Austrian engineer Hans Hörbinger, which asserted that ice is the basic ingredient of the universe. Lenard's musings on racial science and the 'spirit of nature' do not really rise above this level – they show that, even by the time of his Nobel award, he had nothing more significant to contribute to science, but had indeed become its opponent.

When, in the 1920s, Einstein began to experience racially motivated criticism and abuse in the German popular and academic

---

* Steiner has been defended against the charge that he held Nazi sympathies, and certainly he does not seem to have been popular with the National Socialists. They were likely, however, to find little cause for complaint in this comment of his: 'Jewry as such has outlived itself for a long time. It does not have the right to exist in the modern life of nations. That it has survived, nevertheless, is a mistake by world history, of which the consequences were bound to come.'

press, Lenard joined in gleefully. At a meeting of the Society of German Scientists and Physicians in Bad Nauheim in September 1920, Einstein and Lenard were pitched head to head in a debate about relativity.

This confrontation followed an attack on Einstein at a public meeting held in Berlin the previous month, allegedly organized by the Working Group of German Scientists for the Preservation of Pure Science. There was in fact no such body, it having been concocted for the purpose by one Paul Weyland, a far-right fantasist without any real scientific training, who deplored Einstein's theory on the sort of 'common sense' grounds that cranks still choose to employ today. Weyland presaged this event with a letter in the Berlin newspaper *Tägliche Rundschau* recycling old accusations that Einstein had plagiarized the insights of other scientists. The meeting itself took place in the capacious Berlin Philharmonic, where Weyland's rant was accompanied by the distribution of anti-Semitic pamphlets and swastika lapel pins.

Weyland had announced that his lecture was the first in a series of twenty that would lay bare the deceptions of relativity. In the event, only one other followed, by the equally anti-Semitic applied physicist Ludwig Glaser (see page 90). The whole shabby affair aroused wide indignation: the letters of support for Einstein that appeared subsequently in the pages of the Berlin press were by no means all from his colleagues. Planck wrote to Einstein characterizing Weyland's assault as 'scarcely believable filth'. He and others feared that such things would drive Einstein to emigrate from Germany.

Einstein did remain in Berlin, but he was evidently unsettled. He went himself to Weyland's meeting and, somewhat against his instincts and with rare misjudgement, he decided to respond publicly to the attack. His letter in the *Berliner Tageblatt* did at least contain a dash of humour to undercut the risk of pomposity, being titled 'My Answer to the Anti-Relativity Theoretical Co. Ltd'. He admitted that the feeble criticisms of his theory did not really warrant a reply, but also pointed out that the real complaint of Weyland and his acolytes was that Einstein was 'a Jew of liberal international bent'. Einstein also mentioned Lenard (who supported Weyland), saying 'I admire Lenard as a master of experimental physics [but] his objections to the general theory of relativity are so superficial that I had not deemed it necessary until now to reply to them in detail.'

The exchange at Bad Nauheim was no more illuminating, and certainly no more conciliatory. After the Berlin affair, this *Einstein Debatte* was widely anticipated, and the hall in which it took place was packed to the galleries, not just with scientists but with journalists and curious onlookers – and Weyland – who must have been thoroughly bored and mystified by the four hours of technical talks that preceded it. Accounts of the debate differ. Some newspapers reported that it was conducted calmly and objectively, but others stated that Planck, who as the society's president was obliged to be the moderator, was forced on several occasions to intervene to prevent hecklers from interrupting Einstein. In any event, neither Einstein nor Lenard was pleased with the outcome. Einstein was highly agitated afterwards – he later admitted his regrets at 'los[ing] myself in such deep humorlessness' – and his wife Elsa seems to have suffered something of a nervous breakdown. For his part, Lenard felt compelled to resign from the DPG in protest at the event, and he fixed a sign outside his office at Heidelberg announcing that the society's members were not welcome within.

Physics for Hitler

Lenard was not the only influential scientist in the anti-Einstein camp. In 1919 the Nobel Prize in Physics was awarded to Johannes Stark for his discovery of the effect of electric fields on the energies of photons emitted from atoms as electrons jump between their quantum orbits.* In an electric field, the energy of an electron in a particular orbit splits into a whole series of different energies: rungs of a new quantized energy ladder. Stark's discovery of this effect was of some importance, since it revealed a further level of quantum granularity in the structure of the atom. Nevertheless, the 1919 award was perhaps one of the Nobel Committee's least auspicious decisions, for it inflated Stark's already ponderous sense of self-importance and entitlement.

Stark's situation was so close to Lenard's that it is no wonder the

---

* Because electrons in atoms do not in fact follow planet-like orbits around the nucleus but are instead distributed in diffuse clouds, their quantum states are more properly called *orbitals*.

two men forged a firm alliance. Like Lenard, Stark was an experimentalist befuddled by the mathematical complexity that had recently entered physics. He was another extreme nationalist whose right-wing views had been hardened by the First World War. He too felt that Einstein had stolen his ideas, this time over the quantum-mechanical description of light-driven chemical reactions. (Stark never in fact fully accepted quantum theory, even though an understanding of the 'Stark effect' depended on it.) And being a mediocrity who struck lucky, he found himself being passed over for academic appointments to which he was convinced he had the best claim. He attributed this to the self-interest of a 'Jewish and pro-Semitic circle' centred on the (decidedly Aryan) Planck and Sommerfeld, the latter being the alleged cabal's 'enterprising business manager'.* This circle included most of Sommerfeld's students, not least Peter Debye, who was given the professorship at Göttingen in 1914 for which Stark had applied. Lenard's and Stark's enemies suggested that their definition of 'Jewish science' was more or less anything that the two physicists could not understand, and that they placed in the 'Jewish cabal' anyone who threatened to outclass them scientifically. But Einstein was undoubtedly perceived as the ringleader of the whole affair.

By 1922 the situation had deteriorated to such a degree that Einstein declined to speak at a session of the Society of German Scientists and Physicians in Leipzig, fearing that his life might be in danger. This wasn't paranoia. In June the Jewish foreign minister of the Weimar government Walther Rathenau, who Einstein knew well, was assassinated in Berlin by two ultra-nationalist army officers. Lenard had refused to lower the flag of his institute at Heidelberg as a mark of respect for the murdered minister, and as a result he had been dragged from his laboratory by an angry mob of students. Lenard narrowly escaped being thrown into the River Neckar, but the distressing experience only deepened his anti-Semitism. When he was reprimanded by the university, he announced his resignation in disgust. He soon withdrew it when he discovered that the shortlist for his replacement

---

* The accusation is all the more risible when one considers that Sommerfeld was himself somewhat prejudiced. He commented to Wilhelm Wien in 1919 that the 'Jewish-political chaos' of the new Weimar Republic was making him 'more and more of an anti-Semite' – the kind of casually bigoted statement that would raise no eyebrows at that time.

consisted of two 'non-Aryans' – James Franck and Gustav Hertz,* who had won the Nobel Prize together in 1925 – and an experimentalist sympathetic to England, Hans Geiger, who had worked with Rutherford in Manchester. In the end Lenard clung on at Heidelberg until 1929, when he was replaced by Walther Bothe. Lenard's colleagues made Bothe's life so miserable, however, that he moved to the Kaiser Wilhelm Institute for Medical Research in Heidelberg. Lenard so dominated the physics institute at Heidelberg that it was named after him in 1935.

Laue spoke on relativity in Einstein's place at the 1922 conference, earning the abiding enmity of the 'Aryan physicists'. His audience was supplied with pamphlets distributed by Stark decrying this 'Jewish theory'.

When, in the following year, the National Socialists took up arms in Munich to openly challenge the complacent decadence of the Weimar government and free Germany from the Jewish stranglehold, Lenard and Stark recognized a kindred spirit and a hope for the future. In May 1924 they wrote an article called 'The Hitler spirit and science'. Hitler and his comrades, they said,

> appear to us as God's gifts from times of old when races were purer, people were greater, and minds were less deluded . . . He is here. He has revealed himself as the Führer of the sincere. We shall follow him.

The Nazi leader noted this pledge of support, and he and Rudolf Hess visited Lenard at home in 1926.

Stark was in fact the author of his own exclusion from the academic community. Slighted by the opposition from his colleagues at Würzburg to his acceptance of a Habilitation thesis from his student Ludwig Glaser – Glaser's study of the optical properties of

---

* Hertz, the nephew of Lenard's mentor Heinrich Hertz, had a Jewish grandfather, which made him non-Aryan according to the 1933 rules. Although his war service exempted him from dismissal at the Berlin Technische Hochschule, he left anyway in 1934 to take up a lucrative offer from the electrical engineering company Siemens, where during the war he worked on the separation of chemical isotopes for nuclear research. As an experimental physicist he was looked on favourably by Stark, an illustration of how the Aryan physicists tended to pick and choose who was and wasn't 'Jewish in spirit'.

porcelain was regarded as mere engineering, not true science – Stark petulantly resigned from his professorship in 1922. He set up a private laboratory in a nearby disused porcelain factory, using the money from his Nobel Prize to fund this industrial venture (which was against the Nobel Foundation's rules). At the same time he channelled his resentment against academia generally and theoretical physics in particular into a book called *The Present Crisis in German Physics*. Glaser, as we saw, had already embraced his mentor's philosophy and became a vocal propagandist of Aryan physics. He was appointed assistant to the undistinguished engineer Wilhelm Müller, Sommerfeld's politically favoured successor at Munich (see page 103). But Glaser was so virulently racist that he became a liability and was subsequently moved out of harm's way to the fringes of the Reich – Poland and then Prague – where he thankfully fades from history.

By the late 1920s Stark's porcelain venture had failed, and he tried to regain an academic post but was repeatedly passed over in favour of more able candidates. When Sommerfeld opposed his application for a professorship at Munich, this confirmed in Stark's mind that Sommerfeld was a spider in the Jewish web.

How Aryans created science

For Stark and Lenard, the canker at the core of German physics was not merely the nepotism of the Jews and their lackeys, nor the obscure theories and unpatriotic internationalism of Einstein. The fundamental problem lay with a foreign and degenerate approach to science itself. The popular notion that science has a universal nature and spirit, they said, is quite wrong. In an article titled 'National Socialism and Science', Stark wrote in 1934 that science, like any other creative activity, 'is conditioned by the spiritual and characterological endowments of its practitioners'. Jews did science differently from true Germans. Echoing Lenard's fantasy, Stark claimed that while Aryans preferred to pursue an experimental physics rooted in tangible reality, the Jews wove webs of abstruse theory disconnected from experience. 'Respect for facts and aptitude for exact observation', he wrote, reside in the Nordic race. The spirit of the German enables him to observe things outside himself exactly as they are, without the interpolation of his own ideas and wishes, and his body does not shrink from the effort which the investigation of nature demands of him. The German's love of nature and his aptitude for natural science are based on this endowment. Thus it is understandable that natural science is overwhelmingly a creation of the Nordic–Germanic blood component of the Aryan peoples.

Just look, Stark implores his readers, at all the great scientists whose portraits are presented in Lenard's *Grosse Naturforscher* (*Great Investigators of Nature*; 1929): nearly all have 'Nordic–Germanic' features (even, apparently, Italians like Galileo).

In contrast, the Jewish spirit in science 'is focused upon its own ego, its own conception, and its self-interest'. The Jew is innately driven to 'mix facts and imputations topsy-turvy in the endeavour to secure the court decision he desires'. Of course, the Jew can imitate the Nordic style to produce occasional noteworthy results, but not 'authentic creative work'. The Jew suppresses facts that don't suit him, and turns theory into dogma. He is a masterly self-publicist, courting and seducing the press and the public – just look at Einstein.

What Germany needs, then, is a truly German, 'Aryan physics' (*Deutsche Physik*) that rejects the overly mathematical fabulations of relativistic physics in favour of a rigorously experimental approach. And in a formula calculated to ingratiate him to the new leaders, Stark adds that

> The scientist . . . does not exist only for himself or even for his science. Rather, in his work he must serve the nation first and foremost. For these reasons, the leading scientific positions in the National Socialist state are to be occupied not by elements alien to the *Volk* but only by nationally conscious German men.

While the Aryan physicists were incapable of mounting a credible assault on Einstein's relativity in scientific terms, *Deutsche Physik* offered a new line of attack: relativity threatened to undermine the very essence of the Germanic world view. Incorrectly claiming that relativity 'sets aside the concept of energy', the Nazi mathematician Bruno

Thüring asserted that in this aspect one can see 'something concerning the soul, world-feeling, attitudes and racial dispositions'. Einstein, he said, is not the successor of Copernicus, Galileo, Kepler (the canonical Nordic–Germanic scientist) and Newton, but their 'determined opponent':

> His theory is not the keystone of a development, but a declaration of total war, waged with the purpose of destroying what lies at the basis of this development, namely, the world view of German man . . . This theory could have blossomed and flourished nowhere else but in the soil of Marxism, whose scientific expression it is, in a manner analogous to that of cubism in the plastic arts and the unmelodies and unharmonic atonality in the music of the last several years ['degenerate science'!]. Thus, in its consequences the theory of relativity appears to be less a scientific than a political problem.

These ideas were noted and initially welcomed by Hitler. 'That which is called the crisis of science', he wrote,

> is nothing more than that the gentlemen are beginning to see on their own how they have gotten on to the wrong track with their objectivity and autonomy. The simple question that precedes every scientific enterprise is: who is it who wants to know something, who is it who wants to orient himself in the world around him? It follows necessarily that there can only be the science of a particular type of humanity and of a particular age. There is very likely a Nordic science, and a National Socialist science, which are bound to be opposed to the Liberal–Jewish science, which, indeed, is no longer fulfilling its function anywhere, but is in the process of nullifying itself.

Such declarations can scarcely leave one with an impression that the Nazis had much sympathy for – or understanding of – true science. But neither should they be read as some kind of official doctrine that guided the Nazi government's policy on scientific research. Frequently, Hitler's grandiose statements – on this or other matters – had as little real influence on the way affairs were conducted at the daily, prosaic level as do the proclamations of the Pope on the dealings of a local Catholic church. Indeed, Hitler purposely maintained a distance between his own views and edicts and their practical implementation. The actual response of the National Socialist authorities to *Deutsche Physik* was not uncritical acceptance but something rather more complex.

Deutsche Physik *under the Nazis*

The anti-Einstein activism of Stark, Lenard and their fellow travellers continued through the early 1930s. In 1931 a hundred scientists and philosophers contributed to a volume denouncing Einstein and his theories. A few supporters, such as Laue and Walther Nernst, defended him publicly against such onslaughts. But typically his champions would stick up for his theories while avoiding the delicate 'political' matter of his Jewishness.

When Hitler became Reich chancellor, the *Deutsche Physiker* must have felt that their moment had come. And so it seemed – at first. Stark was made president of the prestigious Physical and Technical Institute of the German Reich (PTR) in Berlin in 1933, giving him new pretensions of power. He announced that the PTR would thenceforth take charge of all German scientific periodicals, and at the meeting of the DPG in Würzburg in September 1933 it seemed to Laue that Stark was trying to anoint himself Führer of all German physics. In his opening address as chairman, Laue publicly challenged the Aryan physicists by making an implicit comparison between the theory of relativity and the condemnation of Galileo's Copernican theory by the Catholic Church. Invoking the (apocryphal) story that Galileo had muttered '*eppur si muove*' ('still [the earth] moves') as he rose after kneeling to hear his sentence, Laue made it clear that Einstein's theory would remain true whatever his detractors might assert.

Here once more, Laue's courage in defying Nazi demagoguery and interference was very rare among the physicists. 'To all of us minor figures', Paul Ewald wrote later, 'the very existence of a man of Laue's stature and bearing was an enormous comfort.' His resistance was not without a certain panache: he was said never to go out of doors without carrying a parcel under each arm, since that gave him an excuse not to give the obligatory Hitler salute in greeting. Laue was one of the very few scientists in prominent positions to move beyond private grumbles and little acts of defiance into open admission of his contempt for the Nazis. And unlike Planck, he came to recognize that scientists could not remain 'apolitical'. In 1933 he was among those who chided Einstein for his activism, warning him that 'political battles call for different methods and purposes from scientific research' and that as a result scientists rarely fared well in that arena. But by and by he saw that one could not simply stand aloof from National Socialism. Indeed, he implied to Einstein that he stayed in Germany only because his loathing of the Nazis made him desperate to see their downfall. 'I hate them so much I must be close to them', he told Einstein during a visit to the United States in 1937. 'I have to go back.' After the war, James Franck said that Laue

> was not a daredevil, blinded against peril by vitality and good nerves; he was rather a sensitive and even a nervous man who never underestimated the risk he ran in opposing Nazidom. He was forced into this line of conduct because he could bear the danger thus incurred better than he could have borne passive acceptance of a government whose immorality and cruelty he despised.

When we hear it said in defence of German physicists that not all men can be heroes, we should bear this remark in mind: it is not a matter of how strong your backbone is, but of how much your personal sense of morality can tolerate.

Thanks in considerable measure to Laue – but perhaps still more to infighting among the National Socialists – Stark's attempt to rule German physics came to nothing. He could, however, at least impose his views on the PTR, where he instigated the Führer principle and sacked all Jews from the advisory committee. The following year he was appointed president of the German Research Foundation, which controlled much of the funding for science, and he promptly withdrew funds for work in theoretical physics. (Because of a shift of political power, Stark fell from grace and was forced to retire from this post two years later, whereupon funds for theoretical physics were restored.)

Prompted by Goebbels' Ministry of Propaganda, in the summer of 1934 Stark wrote to all eleven of his fellow Nobel laureates in Germany asking them to sign a letter declaring that

In Adolf Hitler we German natural researchers perceive and admire the saviour and leader of the German people. Under his protection and encouragement, our scientific work will serve the German people and increase German esteem in the world.

This quasi-religious statement found no takers, although the refusals were carefully crafted. Heisenberg, for example, told Stark that he agreed with the sentiments but felt it inappropriate for scientists to make public pronouncements on political matters. That was not just a convenient excuse but a genuine statement of belief, which cut both ways: Heisenberg seemed to apply it equally to Stark's infantile gesture and to questions of moral responsibility.

Stark and Lenard fretted about the KWG, which seemed to them to be decidedly lax about expelling its Jewish members – no doubt, they were convinced, because it was dominated by an Einsteinian cabal. 'From the beginning', Lenard wrote in 1936, 'it was . . . a Jewish monstrosity with the purpose, entirely unknown to the emperor and his advisers, of enabling Jews to buy themselves respectability and of bringing Jews and their friends and similar spirits into comfortable and influential positions as "researchers".' Starting now to ramble inanely, Lenard proclaimed that the society's president Planck was 'so ignorant about race that he took Einstein to be a real German', doubtless because of the many theologians and pastors in Planck's family and their misguided respect for the Old Testament.

Stark and Lenard had hoped to set the society straight when Planck's first term of office came to an end in 1933: 'to make something sensible of this completely Jewish business', wrote Stark, 'which, as a start, must simply be pulled to pieces'. But Planck did not retire; he stayed for a second term of office. When that was due to expire in March 1936, Stark felt sure he would be called upon as the new president. Inexplicably, he wasn't. (Bernhard Rust, who was now able to dictate the society's affairs at the Reich Education Ministry, distrusted Stark, who had aligned himself with Rust's political opponents in Nazi circles.) Well then, said Stark, it must be Lenard. Rust approved of that idea, but now Lenard himself declined, saying he was too old. No other successor was put forward, and meanwhile Planck stayed on.

It was a delicate moment, since the Aryan physicists weren't alone in regarding the KWG as ideologically suspect. After the society's twenty-fifth anniversary celebrations in January 1936, the Nazi newspaper *Völkischer Beobachter* called it a 'playground for Catholics, Socialists and Jews', while the SS journal *Das Schwarze Korps* had portrayed it as a 'restricted circle' basking in elitist 'aristocratic splendour'. Planck knew that Rust would not endorse a replacement who was too closely associated with Einstein, and would prefer someone known to be faithful to the party. The minister would also insist that the organization now adopt the Führer principle. But the KWG senate cannily identified a candidate who, as an industrialist, could retain some independence from political influence, while as a staunch patriot should be unobjectionable to the leaders: the chemistry Nobel laureate Carl Bosch. He was duly elected in 1937. But in place of the secretary Friedrich Glum, Rust appointed the Nazi official Ernst Telschow, who had some chemical training and had worked briefly under Otto Hahn. As Bosch was frequently plagued by illness, Telschow took over much of the society's practical business. Arguably this was no bad thing for the KWG, for Telschow was a canny administrator, able to form links with the Nazi regime that would benefit the society. One of those individuals who knew how to adapt to the prevailing political climate, Telschow was active in the (renamed) society after the war and was finally elected a senator in 1967.

While the KWG was not exactly Nazified in 1937, then, neither did it thenceforth mount any effective resistance to the wishes of the government. It expelled the remaining Jewish members, including Lise Meitner, even though she continued to work at Hahn's institute in Berlin.

White Jews

This outcome did not afford the *Deutsche Physiker* much satisfaction, and in 1937 Stark decided it was time to find another line of assault on his enemies in theoretical physics. Planck's influence was evidently waning, and now Stark found a new target: a young professor who was enjoying the fame that Stark so coveted and who had made quantum theory an even more impenetrable thicket of mathematical formalism, who supported Einstein's ideas, had been awarded a Nobel Prize at the absurdly premature age of thirty-one, and now looked about to be appointed as Sommerfeld's successor in Munich. Stark began a crusade against Werner Heisenberg.

Heisenberg had been in Stark's sights ever since he had refused to attend the rally of the National Socialist Teachers League in Leipzig in November 1933. On that occasion Stark hoped to agitate Heisenberg's students into protest, but Heisenberg defused the situation by inviting the leader of the local Nazi Students League to his house and persuading him that he was a trustworthy, albeit 'apolitical', professor. Emboldened by this victory, at the gathering of the Society of German Scientists and Physicians in Hanover in September 1934 Heisenberg defended relativity and quantum theory against Stark's accusations that they were speculative. There he even mentioned Einstein by name, earning him a reprimand from the Nazi chief ideologue Alfred Rosenberg.

But by 1935 Heisenberg was deeply disheartened by the political climate. His sense of patriotism and honour was disturbed after the Nuremberg Laws had removed the exemption from dismissal for Jewish veterans of the First World War. He had even risked damaging his reputation and prospects by registering that displeasure at a faculty meeting. His words of protest, however, show how the Nazis had already set the parameters of the debate: Heisenberg said he doubted 'that the measures now being taken are consistent with the intention of the law, according to which front veterans also belong to the *Volk* community'. In other words, it was not the principle of an exclusive national community that he challenged, but who was selected for membership.

On that occasion Heisenberg had considered resigning (or so he claimed), but was dissuaded by Planck, who cautioned once again that this would be a futile dereliction of duty. 'It is to the future that all of us must now look', the older man advised: they must hang on regardless, for Germany's sake. Like most of his peers, Heisenberg withdrew into physics. 'The world out there is really ugly', he wrote to his mother, 'but the work is beautiful.'

The immediate trigger for Stark's attack on Heisenberg in 1937 was a long-running dispute about the successor of Arnold Sommerfeld, who two years earlier had been due to retire from his professorship in Munich. It was no secret that Sommerfeld wanted Heisenberg to have the post, and it was said that the 'list' of candidates submitted by the university to the Bavarian administration contained his name and no other.

Stark and Lenard hoped that Sommerfeld's departure could be used to free the Munich faculty from his baleful support of 'Jewish physics'. In an address at the new Philipp Lenard Institute for Physics in Heidelberg in December 1935, Stark called Heisenberg a 'spirit of Einstein's spirit'. This speech was printed in the January issue of the party periodical *Nationalsozialistische Monatshefte*. In February Heisenberg placed a response in the *Völkischer Beobachter*, although it was printed with a further comment from Stark. Concerned about the damage to his career and reputation, Heisenberg sought an audience with Rudolf Mentzel, Rust's deputy at the REM, at which he argued that theoretical physics was important and needed to be defended against the diatribes of the *Deutsche Physiker*. Probably because of internal party politics rather than scientific judgement, Mentzel looked favourably on the appeal, but advised Heisenberg to send a letter to all German university physics professors asking if they took the same view. Together with Max Wien, a physicist at Jena, and Hans Geiger – both carefully selected as experimentalists sympathetic to his cause – Heisenberg drafted the letter, which demanded that the attacks of Stark and Lenard should cease for the sake of Germany's international reputation. Nearly all of the seventy-five professors who received the letter signed their approval.

Thus Stark had succeeded only in showing the REM that there was scarcely anyone else on his side. To make matters worse, he was forced to resign as head of the German Research Association in November 1936 after squandering its funds on a hare-brained idea to extract gold from the moors of southern Germany. But this apparent victory did little to improve Heisenberg's mood. Despite marrying in early 1937, he found himself mired in despair and gloom in Leipzig, apparently close to a breakdown and admitting that, when he was not with his new bride, 'I now easily fall into a very strange state.' In March he was finally offered Sommerfeld's professorship, which he accepted but deferred until August. That turned out to be a mistake, because it gave Stark the chance to intervene again.

In July Stark published in *Das Schwarze Korps* a new, trenchant vilification of Heisenberg, along with others who colluded in the 'Jewish conspiracy' in physics without being Jews themselves. These people, he said, were 'White Jews' – a designation calculated to make them the legitimate targets of all the abuse previously heaped on the Jews themselves. Planck, Sommerfeld and their circle were denounced as 'bacterial carriers' of the Jewish spirit who 'must all be eliminated just as the Jews themselves'. And none more so than Heisenberg, 'this puppet of the Einsteinian "spirit" in new [Weimar] Germany'. Even today, Stark claimed, the core of Heisenberg's students 'still consists of Jews and foreigners'. The young pretender himself was the 'Ossietzky of physics', implying that he was no less dangerous to German culture than the dissident Carl von Ossietzky who the previous year had been awarded the Nobel Peace Prize (see page 121) – and that Heisenberg, like Ossietzky, should therefore be in a concentration camp. A disgusted Peter Debye showed the article to the senate of the KWG, reporting that 'it was condemned by everyone with whom I spoke'.

Heisenberg was now in a bind. He had to extricate himself from the 'White Jew' accusation without appearing to distance himself from Einstein's 'Jewish' physics. His response was telling: it was not enough simply to defend his good character, he also sought official sanction from the state leaders. Thus he directed his appeal to the Reichsführer of the SS, Heinrich Himmler, insisting that he must either have complete vindication at the highest level or he would resign and emigrate. He reminded the authorities that he had plenty of offers from abroad, in particular from Columbia University in New York. Having previously refused to 'desert' Germany in the face of the Nazi excesses, he thus contemplated it, or at least threatened it, now to save his 'honour'. As historian Paul Lawrence Rose argues, Heisenberg's counter-attack on Stark should not be interpreted as a rejection of Nazism or anti-Semitism; it was driven by pride, anger, and fear for his reputation.

In cases like this, one needed to exploit personal connections for all they were worth. Heisenberg's mother was acquainted with Himmler's mother, and she argued her son's good character in a way that Frau Himmler would appreciate: as mother to mother. Frau Himmler promised that she would get her Heinrich to 'set the matter back in order'. 'There are some slightly unpleasant people around Heinrich', she admitted, 'but this is of course quite disgusting. He is such a nice boy – always congratulates me on my birthday.'

Himmler, however, at first remained neutral. He simply requested a detailed response from Heisenberg to the accusations made by Stark, while at the same time ordering an investigation into Heisenberg's character. The Gestapo and SS bugged Heisenberg's house, placed spies in his classes, and questioned him on several occasions. This exhausting and frightening process finally resulted in a report that exonerated Heisenberg, portraying him as an 'apolitical' scientist who was basically a good patriot with a positive attitude towards National Socialism. It explained that Heisenberg had initially been trained in 'Jewish physics', but claimed that his work had become increasingly 'Aryan'. True, he did not show the antipathy towards Jews that one might hope for, but perhaps he would develop the proper attitude in due course.

Himmler received the report in the spring of 1938, but to Heisenberg's immense frustration he did not act at once. Finally in July he was prevailed upon to write to Heisenberg, saying 'I do not approve of the attack of *Das Schwarze Korps* in its article, and I have proscribed any further attack against you.' He invited Heisenberg to discuss the matter with him 'man to man' in Berlin later in the year. The invitation was, despite Heisenberg's eagerness, never fulfilled, but the two men remained in cordial contact through the war. Given the other demands on Himmler's time, the attention he gave to this matter is in fact rather remarkable. Mark Walker attests that Himmler was very interested in science and considered himself something of a patron of scientists. A personal letter and invitation from Himmler was more than most of them might have expected.

It was nonetheless a ruthless kind of patronage. When Himmler explained his decision on Heisenberg to the head of the Gestapo Reinhard Heydrich, he wrote with icy pragmatism that 'I believe that Heisenberg is decent; and we cannot afford to lose this man or have him killed, since he is a relatively young man and can bring up the next generation.' Moreover, Himmler concluded with a bathetic indication of his scientific ignorance, 'we may be able to get this man, who is a good scientist, to cooperate with our people on the cosmic-ice theory'. To Heisenberg's good fortune, it seems he was never asked to give an opinion on the matter.

Himmler also added chilling words of advice in his letter of exoneration to Heisenberg, saying 'I would consider it proper, however, if in the future you make a clear distinction for your listeners between the recognition of the results of scholarly research and the personal and political attitude of the researcher.' In other words, Heisenberg would do well not to mention Einstein. He got the point, and obeyed.* He had already indicated that intention in a letter sent in March to Ludwig Prandtl, an expert in aerodynamics at Göttingen, who had tipped off Heisenberg that exoneration from Himmler was on its way:

> I never was sympathetic toward Einstein's public conduct . . . I will gladly follow Himmler's advice and, when I speak about the theory of relativity, simultaneously emphasize that I do not share Einstein's politics and world view.

Having been granted his wish to 'set the record straight' with the guarantee of an article in *Zeitschrift für die gesamte Naturwissenschaft*, the house journal of the *Deutsche Physik* movement, he pursued this concession doggedly over the next few years, again asking Himmler to intercede when difficulties arose. That his article, 'Evaluation of the "modern theoretical physics"', was not actually published until 1943 rather defeated its original object. He consented therein to the usual compromise of acknowledging Einstein's discoveries while suggesting that they would have happened anyway:

> America would have been discovered if Columbus had never lived, and so too the theory of electrical phenomena without Maxwell and of electrical waves without Maxwell, for the things themselves could not have been changed by the discoverers. So too undoubtedly relativity theory would have emerged without Einstein.

---

* In 1942 Sommerfeld was about to publish some lectures on physics when he received a letter from Heisenberg saying (as Rudolf Peierls later recalled it) that 'a political adviser and close friend of mine, also a physicist, would like to call to your attention certain guidelines which are now in use, that is, we note, the publisher noticed that you mentioned Einstein's name four times in your lectures, and we wondered if you couldn't get by with mentioning him a little less often?' Sommerfeld complied, retaining just one of the references. 'I must mention him once', his conscience obliged him to write back. Peierls adds that 'after the war the names were quickly put back in'.

These accommodations and entreaties to the Nazis may seem hard to understand today. Could Heisenberg really have imagined, after an episode like Stark's attack, that things were going to get any better? That, if he could only 'clear his name', somehow the relationship of physics with the National Socialist state could be set back on track? But it was not naïve optimism that kept him bound to the Fatherland, but rather, 'an unbreakable attachment to Germany [that] his entire life and upbringing had instilled in him', as his biographer David Cassidy puts it. To Heisenberg, Cassidy says, 'remaining in Germany was apparently worth almost any price, as long as he could continue to work and teach'. What is more, Heisenberg had developed a conviction that his own fate was tied to that of the whole of German physics; if he left, nothing would remain. But as Cassidy points out, 'by seeing himself in such a grandiose rationalization for remaining in Germany, he more easily succumbed to further compromises and ingratiation with the regime'.

In fact things really did improve eventually for Heisenberg, if not necessarily for German physics: by 1944 he was celebrated in Goebbels' weekly propaganda newspaper *Das Reich* as a 'German national leader'. This only lends weight to Rose's accusation that 'Heisenberg's notion of "responsibility" as the acquisition of influence in Nazi circles was actually a rationalization of collaboration and of self-interest.'

What of the Munich post that had prompted Stark's assault? In that regard Stark was indirectly successful, preventing Heisenberg ever from becoming Sommerfeld's heir. The position fell foul of political wrangling between the REM, the SS, the Munich faculty and the Nazified University Teachers League, out of which Sommerfeld's replacement emerged on the eve of war in 1939, in the form of an undistinguished mechanical engineer named Wilhelm Müller, who opposed the 'new' physics and would teach only the classical variety. When Walther Gerlach, an expert in quantum theory at Munich, complained to the dean of the university that no theoretical physics was now being taught there, he was curtly told that

If you only understand theoretical physics to mean the so-called modern dogmatic theoretical physics of the Einstein–Sommerfeld stamp, then I must inform you that this will indeed no longer be taught at Munich.

The wrong battle?

The battle fought within German physics in the 1930s was not that of apolitical scientists against the National Socialists, but of Einstein's supporters against *Deutsche Physik*. One might have expected the National Socialists to embrace a view of physics that discredited Jews, but they were not quite as foolish as that. Physics under the Nazis was never really hijacked by ideology, for the political leaders were primarily interested in practical outcomes and not academic disputes. An internal REM memo to Bernhard Rust on the controversy over 'Jewish physics', probably sent by the ministry's undersecretary (who here seems concerned that the blundering Rust might make a fool of himself), advised that 'In the case of a purely scientific dispute, in my opinion, the minister should keep himself out of it.' Until nuclear fission was discovered in 1938, the new theoretical physics was of little interest to the authorities, as it seemed to be largely irrelevant to the war preparations. And once atomic power looked possible, it was clear that the Aryan physicists' advocacy of practical experiment over abstract theory could not deliver results. Rather, it was evidently the proponents of 'Jewish' quantum theory and relativity who truly understood the secrets of the atomic nucleus, and even the Nazis could see that they were the only ones likely to put the discoveries to good use.

Deutsche Physik also floundered through the political ineptitude of Stark and Lenard. Stark in particular was apt more to antagonize than to persuade the party officials. 'Had he been less crazy', science historian John Heilbron comments laconically, 'he would have been much more dangerous.' The Aryan physicists made wild blunders, but more incapacitating was their failure to appreciate that to get your way in Nazi Germany you needed to do more than regurgitate approved doctrines, prejudices and formulas. You needed to be able to manipulate the competing power blocs, to exploit the right contacts and forge useful alliances. Stark often backed the wrong horse – he had no more judgement in politics than he did in science.

As a result, the attempt of *Deutsche Physik* to take over the academic system failed. But its opponents had to tread a fine line, so that their defence of Einstein's theories did not risk endorsing his unpopular political views. So long as they agreed to avoid too explicit an acknowledgement of the architect of the theory of relativity, they could generally get their way. During the war Heisenberg regularly omitted Einstein's name from the public lectures that he was asked to deliver to spread German culture in occupied territories. Indeed, historians Monika Renneberg and Mark Walker suggest that *Deutsche Physik* collapsed partly because it was rendered otiose by the compromises made by the mainstream physics community, which demonstrated, to the leaders' eventual satisfaction, 'their willingness and ability to help further the goals of National Socialism'.

The struggle against *Deutsche Physik*, although frustrating for the German physicists who rejected it, offered a convenient narrative after the war by supplying a criterion for partitioning physicists into those who were Nazified and those who resisted them. In this view, if you had opposed Aryan physics, you had in effect opposed the Nazis – all the guilt of the National Socialist era could be transferred on to Lenard, Stark and their supporters. Better still, one could use this division to apportion scientific competence: the Aryan physicists were universally poor scientists, their opponents always proficient.

But the truth was that, while the dispute rumbled on through the late 1930s, the Nazis tightened their grip on German science regardless. In some disciplines, such as chemistry, scientists fell into line in short order. In a few, such as anthropology and medicine, the collusion of some researchers had horrific consequences. Physics was another matter: just docile enough for its lapses, evasions and occasional defiance to be tolerated. The physicists were errant children: grumbling, arguing among themselves, slow to obey and somewhat lazy in their compliance, but in the final analysis obliging and dutiful enough. If they lacked ideological fervour, the Nazis were pragmatic enough to turn a blind eye. Their attitude is conveyed perfectly in a description of Ludwig Prandtl sent by the local Nazi coordinator (*Kreisleiter*) in Göttingen to his superiors in May 1937. As we saw, Prandtl had supported Heisenberg against Stark's attacks, and he had appealed to Himmler about the damaging effects on German science of the *Deutsche Physiker* attacks. The *Kreisleiter*'s letter makes it clear how indifferent the Nazis were to such arguments, and how meaningless or even contemptible the notion of a 'duty to science' was to them. All that mattered was whether the scientists were prepared to lend their efforts to mobilization of the Fatherland, which Prandtl did willingly:

> Prof. Prandtl is a typical scientist in an ivory tower. He is only interested in his scientific research which has made him world famous. Politically, he poses no threat whatsoever . . . Prandtl may be considered one of those honourable, conscientious scholars of a bygone era, conscious of his integrity and respectability, whom we certainly cannot afford to do without, nor should we wish to, in light of his immensely valuable contributions to the development of the air force.

7 'You obviously cannot swim against the tide'

In 1936 Peter Debye, now director of the Kaiser Wilhelm Institute for Physics housed in its own premises in Berlin, decided that the newly constructed institute should be renamed in honour of the venerable colossus of German science, Max Planck. He anticipated resistance, both because traditionalists would be loath to disregard the imperial past and because Planck was considered politically suspect by the Nazi regime. Debye's characteristic strategy was to make the decision a fait accompli by having the new name – the Max Planck Institute for Theoretical and Experimental Physics – carved in stone above the institute's entrance, ingenuously claiming that he simply wanted to give Planck a pleasant surprise on the institute's inauguration day. It is said that when the Nazis predictably ordered him to remove it, Debye instead covered it with a wooden plank – the pun works in German too.

That was Peter Debye: boldly and wittily outflanking his opponents while shrugging off political attempts to control and manipulate his science. At least, so the story suggests. But there is no first-hand record of the 'plank' incident, and it is quite possibly no different to so many other tales in science history, retold for the sake of its lustre without regard to documentary evidence. All the same, Debye did rename his institute, and thereby set in train a process by which eventually the entire KWG, the research network at the heart of German science, became associated with the foundational role of Max Planck. It is today the Max-Planck-Gesellschaft, and its research centres are all Max Planck Institutes.

The clearest account of the renaming of the KWIP is given by the Rockefeller's Warren Weaver during a visit to Berlin in January 1938. It corroborates Debye's resolve, but with less swagger:

We visit Debye's institute, which has been essentially completed now for a few months. It has still not been formally dedicated, there being official trouble in connection with the name of the institute. On the outside of the building, over the front entrance, one finds the name – The Planck Institute of Theoretical and Experimental Physics – but inside in the entrance hall the plaque which bears an inscription to Planck is covered up with a cloth. Stark and Lenard both wrote letters to the minister [Rust] insisting that Planck was not a great enough physicist to warrant to name the institute after him. Debye has discussed this with the minister. He simply says Debye must be a little patient, and things will adjust themselves, although he cannot at the moment insist on a move which is opposed by strong party members. D. is completely untroubled by these circumstances, saying that the institute is open to scientific research, which is his only concern.

Debye got his way. When the KWIP was eventually dedicated on 30 May 1938, it was as the Max Planck Institute – although this name was delicately omitted from the official invitation. To acknowledge the joint sponsors of the project, the entrance was adorned for the ceremony with two flags, bearing the stars and stripes and the swastika.

Debye's critics today have little regard for this 'victory', instead seeing in his role as KWIP director just another instance of his will-ingness to take positions of influence under a regime that no one could now fail to see as totalitarian, racist, corrupt and warmongering. That perspective sheds little light by itself. The real question is how Debye, who evidently would have preferred the Nazis never to cross his threshold, reconciled himself to the compromises that this entailed. The period that Debye spent as head of the KWIP – during which he became ever more central to German physics – and the circumstances that terminated this position are critical to the matter of how posterity should regard his moral conduct.

The virtual institute

When Debye was appointed director of the KWIP in 1934, there was nothing concrete for him to administer. Ever since its inception in 1914 the physics institute had existed only on paper, being little more than a mechanism for dispensing grants. The KWG had included a physics institute in its original plans, but the case for establishing it in Berlin was not strong, for the city already hosted the renowned Physical and Technical Institute. All the same, in January 1914 Planck, Haber and Walther Nernst persuaded the Prussian Ministry of Culture to lodge an application for building premises there. 'The purpose of this institute', that document proclaimed,

> will be to solve important and urgent problems in physics, and secondly to form associations of physical scientists specifically suited to the issues involved . . . The site of the institute should be in a small building in Dahlem, which provides the opportunity for meetings and for hosting archives, a library, and physical equipment.

Einstein was proposed as the institute's first director. The application was rejected by the finance minister on the day before the outbreak of the First World War. Surprisingly, it was revived in the midst of the war, despite there being no suggestion that the institute would be involved in military research. At this stage there was no longer talk of a building, however: the 'institute' would consist simply of a board of trustees who would allocate funds for research conducted elsewhere, making it in effect a grant-giving agency. Einstein was made director, and the board included Planck, Haber and Nernst along with representatives of the various governmental and industrial sponsors, including Friedrich Glum, a member of the Prussian Interior Ministry who became director general of the KWG in 1922. In 1921 Max von Laue was elected to the board and was soon thereafter appointed deputy chairman, taking over much of the administration as Einstein's interest in the institute waned.

For a time this arrangement worked well enough. For example, in 1918 the KWIP awarded Debye and his assistant Paul Scherrer in Göttingen money to buy new X-ray equipment. (The disruptions of the post-war situation meant that he didn't get it until the summer of 1920, by which time he was in Zurich.) But by the late 1920s hyperinflation had severely depleted the KWIP's financial resources, and it became clear that it would have rather little significance unless it were able to function as a centre of research itself. In 1929 the institute's committee, probably motivated by Laue, appealed for a building. But who could pay for it?

The biochemist Otto Warburg, whose physicist father Emil was on the KWIP board, had an answer. On a lecture tour of the United States during that year he had made contact with the philanthropic Rockefeller Foundation. The foundation had already helped to fund the KWG's activities earlier in the decade (see page 20), including the building of an institute for psychiatry in Berlin-Dahlem that opened in 1928. Warburg secured an agreement from the foundation to pay for an institute for cell physiology, and now the KWIP committee added a request for funds for a physics institute too.

In February 1930 the Rockefeller Foundation sent Lauder Jones, its representative in Paris, to Berlin to consider the application. The institute, he reported back in what was intended as a recommendation, would be 'erected primarily for Einstein and von Laue'. The Rockefeller's management was sympathetic but wary of being saddled with an indefinite financial commitment, and wanted assurance that the state would be able to take over funding once the institute was up and running. Glum's attempts to secure that commitment from the Weimar government were thwarted by bureaucracy, but nevertheless in April the Rockefeller's administration approved a grant of $655,000 towards 'land, buildings and equipment' for the Kaiser Wilhelm Institutes for Cell Physiology, led by Warburg, and for Physics, led by Einstein and Laue. The physics institute would conduct experiments on molecular rays and magnetism, and would also have a theoretical division. In view of Einstein's diminishing involvement, it was agreed that Laue would be the 'active' director on the site; Glum admitted to Jones in January 1931 that it was no longer clear if Einstein would wish to be affiliated at all, or whether he would 'prefer to stay in his own home to think'.

By March it had become apparent that thinking was more attractive to the architect of relativity than trying to raise a building, and Planck began looking for a new director. He considered the experimentalist Hans Geiger, but his preferred choice was Nobel laureate James Franck at Göttingen. Nernst was about to retire from his position at the University of Berlin, and if Franck could fill the vacancy, he would be conveniently located to act as director of the KWIP too.

As the financial and political disturbances in Germany grew more severe, Glum was forced to admit to Jones that the building would have to be postponed. Matters had barely progressed by the time

Germany acquired a new government in January 1933 and descended quickly into dictatorship. Suddenly, providing funds for German science looked rather more fraught from the other side of the Atlantic. But promises were promises, weren't they?

Funding Hitler

In late 1936 a reporter from the *New York Times* turned up unannounced at the offices of the Rockefeller Foundation, demanding a statement about the 'large gift' that the foundation had apparently made to the 'Hitler government'. The Rockefeller staff told him that they 'attempt to apply uniform and objective criteria to our projects, making no distinctions of country, race, creed, or politics'. But they were evidently discomforted, for of course such distinctions were precisely what the Hitler government had imposed in Germany.

The Rockefeller's funding of the two Kaiser Wilhelm Institutes in Berlin threatened to become something of a scandal. The *New York Times* ran its story on 24 November under the headline 'Rockefeller Gift Aids Reich Science'. The report quoted the foundation's president Raymond Fosdick as saying that 'The world of science is a world without flags or frontiers', while at the same time he admitted 'it is quite possible that the foundation would not have made the grant if it could have foreseen present conditions in Germany'. In response, Felix Frankfurter, a professor of law at Harvard and adviser to Franklin D. Roosevelt, wrote to Fosdick personally to say that by giving money to the Nazis the foundation had 'adulterate[d] the spiritual coinage of the world'. Fosdick could only reiterate that, as the *New York Times* had stated, this was not a gift to Hitler but merely the fulfilment of a pledge made in 1930.

All the same, the Rockefeller directors in New York were anxious to know just how their money was being spent in Berlin. The officers on the ground, especially Jones, Wilbur Tisdale and Warren Weaver, had misgivings from the moment the Nazis came to power and had been keeping a wary eye on how matters were evolving. According to Alan Gregg, another Rockefeller 'scout' in the Paris office, Warburg said in October 1933 that he felt 'the Nazi regime will not slacken its interest in the development of scientific institutes in spite of its anti-Semitic activities'. The following June, Warburg assured Tisdale that 'less ignorant and more moderate forces are gaining ground'. But by July, Thomas B. Appleget in Paris admitted to his boss Max Mason in New York that he was worried about 'the attitude of the present and future German governments to pure science'. The Berlin chemical institute, he said, is 'now given over entirely to work in the field of chemical warfare; the institute in Munich [psychiatry] is almost entirely dominated by projects in the field of "race purification"'. The Nazis had appointed the Swiss-born eugenicist Ernst Rüdin as head of the Munich institute, where he advanced racial theories that endorsed National Socialist policy. 'What', Appleget added ominously, 'might the physics institute be in five years?'

But even if the Germans were discriminating against the Jews, Tisdale argued, wouldn't it just compound the offence if the Rocke-feller were 'to follow their example and refuse opportunity to the Germans because they are Germans'? Besides, there were grounds for optimism. In July 1934 Planck conveyed to Mason news that he hoped would be reassuring: the KWIP had now confirmed its new director, the staunchly apolitical Peter Debye.

James Franck's departure from Göttingen and emigration to the United States in 1933 had stymied Planck's original hope of making him the director, even though he initially regarded the move with complacency, telling Tisdale that he felt Franck would be able to return 'within a year or two'.* By November of that year Planck had become more realistic and selected Debye as a replacement. At that time, the president of the Physical and Technical Institute in Berlin was Johannes Stark, who was predictably opposed to any choice backed by Planck. When in May 1934 he heard of Debye's likely appointment, he wrote to the minister of the interior alleging quite untruthfully that Debye lacked experimental skills:

Professor Debye is in my opinion not suitable to head an institute for nuclear research. He is an outspoken theorist and as such, is dependent

* Franck settled at the University of Chicago and played a prominent role in the Manhattan Project. He chaired the committee on Political and Social Problems relating to the atomic bombs, and as such he oversaw the Franck Report in June 1945 which recommended that the bombs not be used on civilian targets but instead demonstrated to other nations in an unpopulated region. That humane advice was, of course, ignored.

for experimental work on the help of experimental physicists. Mr Planck is a pure theorist and does not know about the requirements for a physical institute.

Stark was, however, somewhat equivocal in his criticism, probably because Debye's work was not as deeply bound up with the relativity and quantum theory that he so detested. He is, Stark assured the minister, 'the best theoretician to be working in a German university today and should therefore be given an appointment that induces and enables him to enlist a school of theoreticians who are better qualified than the formalists Einstein, Schrödinger, Heisenberg, etc. to promote the progress of physics instead of impeding it'. It looks almost as if Stark suspected (wrongly) that Debye might be turned into a useful ally.

Debye had his own misgivings about Planck's offer, fearing that the proposed simultaneous appointment at the University of Berlin would commit him to too much teaching. Besides, he told Tisdale in June 1934, he had heard that the contract for directors of Kaiser Wilhelm Institutes was not secure, but rather 'an arrangement of convenience and can be broken by the state at will'. And he didn't want to be caught out as he had been in Göttingen, unwittingly jeopardizing his Dutch citizenship by accepting an academic position in Germany. Neither did he relish the thought of another struggle like the one he had endured in coming to Leipzig, when, obliged like all German academics to swear an oath of allegiance to Hitler in front of the dean, he had added by hand to the written declaration, 'Given on the understanding that it will not affect my citizenship.' To ensure that an appointment to Berlin would not compromise his nationality, he appealed at the highest levels. His case was brought before Holland's Queen Wilhelmina, who renewed Debye's citizenship and granted him permission to take the Berlin post. Meanwhile, Debye insisted to Bernhard Rust that it must be a condition of his move to Berlin that the usual German law on transfer of citizenship be waived.

Why, if Debye was intent on pursuing a career in Germany, did his Dutch identity matter to him so much? It's not as though he felt any strong affinity to or patriotism for the country of his birth: like many natives of Maastricht, he identified more with the polyglot character of Limoges than with the culture of Flanders and Holland. Debye was no doubt keenly aware of the special status he acquired as a non-German in a position of authority, which gave him more room to manoeuvre than a German national would enjoy. 'Being a Dutch citizen at this time was extremely important', his son Peter later explained, 'because it gave us a certain immunity to everyday pressures that the German citizens were experiencing. This kind of isolation made our life much easier and kept us away from the anxiety and fear that forced people to refrain from any loud derogatory remarks [about the Nazis] anywhere in public.' As a memo of the Rockefeller Foundation observed in October 1939, Debye's 'outsider' position in Germany gave him 'something like a diplomatic status'.

Despite his hesitation, Debye seems already by April 1934 to have made up his mind. He was approached that spring by Sommerfeld, who wondered whether Debye might consider becoming his successor in Munich. Debye replied:

> I love Munich, and your presence, together with the small laboratory where I could develop some experimental ideas with good people, is indeed attractive. But as things stand, I need to keep faithful to Planck and hope that I will be able to use my strength in the way he has planned.

In July, Debye – on a visit to Liège in Belgium until the following April – formally accepted the KWIP post. This meant that he was also accepting the chair vacated by Nernst at the University of Berlin. But disputes about that post continued for over a year, so Debye (possibly to his relief) did not play any role at the university until March 1936, and did not deliver a lecture until the winter term.

Debye had already impressed the Rockefeller agents as a forthright individual who could resist Nazi interference. When Warren Weaver visited him in Leipzig in May 1933, he heard how a student had challenged Debye openly about the idea of a foreigner holding a prestigious professorship in Germany. Debye had retorted that he did not discuss such questions with a 'little man', but that if a 'big man' came up and expressed the same opinions, he would leave at once. This emboldened attitude of nationalistic students towards professors who would once have been shown absolute deference was a sign of things to come, and one did not brush aside such complaints without risking the wrath of the Nazified student organizations. Yet Debye seemed hardly more perturbed when the 'big men' did intervene. When the Leipzig authorities questioned his selection of an assistant on the grounds solely of scientific merit rather than of political persuasion, Debye replied that 'he would take a page from the book of the Führer and would be a dictator in his own laboratory'. This was a bold double bluff – while it sounds as though he is mocking Hitler, Debye knew that the Nazis wanted to see the Führer principle applied at all levels of society. His remark *was* interpreted by some as disrespectful, but Debye rode out the repercussions, and when Tisdale visited him in June 1934 he reported that 'Debye seems to stand more firmly than ever because of his display of backbone.'

But Tisdale was by no means sure that Debye could retain such autonomy at the KWIP, telling Weaver at the start of August that 'under the present regime I have no confidence in the belief that he would have a free hand, nor be free from abuses imposed by the incompetence or worse of the present regime'. At this stage the Rockefeller Foundation was still wondering if it ought to continue with its pledge to the KWG.

Planck was alarmed. He assured Tisdale that at the KWIP Debye alone would have 'the power to decide on the selection of his co-workers, and to this extent the freedom of scientific research at the institute will be guaranteed in the most complete manner conceivable'. It was what Planck desperately wanted to believe, not just to secure the American funding but also because the physics institute had taken on a symbolic status for him. It was here, in a place relatively insulated from the manipulations of the Nazis, that German physics could be preserved until better times – which he was sure would not be so long in coming. For Planck the KWIP had become an ark that would rescue them from the deluge, and Debye the captain who would steer the ship into a safe port.

Tisdale remained sceptical. Although he agreed with Planck that Germany needed a first-class physics institute, he admitted that 'the appeal leaves me quite cold when I realize that because of the race prejudices they have exiled some of the very men who could have given them the physics which they now claim they so much need'. Planck was also having difficulty in persuading the Rockefeller Foundation that the Reich would keep its side of the funding bargain, as promised by Glum during the Weimar era. When Tisdale asked Planck in July for written confirmation that the German government would provide the 100,000 marks it had pledged, he admitted that 'negotiations are interminably slow and met at every step by indecision and red tape'. Earlier that year he had made a direct appeal to Goebbels, reminding him what the institute might achieve under a man like Debye:

> There is no doubt that under his leadership, the institute, particularly in the field of atomic physics, would open up new areas of science, of which no one can tell in advance whether they might not, like the wireless waves or X-rays also discovered by German physics professors through purely scientific laboratory work, bring about revolutionary changes in public life.

Planck had a shrewd idea of the propaganda minister's priorities. If the delays persisted, he warned, and the Rockefeller Foundation decided to withdraw their support,

> an opportunity would be missed [to] build a plant which would benefit German science and the whole country and which would also be most effective in quelling international talk of the lack of understanding of the new government towards the maintenance of scientific research.

Building the ark

Whether or not Goebbels was swayed by this rhetoric isn't clear, but in early 1935 Planck finally secured a promise that the Reich would provide ongoing financial support for the KWIP. In February he wrote to Debye in Belgium to say that construction of the institute could begin. It was to be located on land long assigned for the purpose next to Warburg's institute in Dahlem, designed by the architect Carl Sattler, who had been responsible for the KWG's Harnack House. Debye took active charge of the project when he returned to Germany, extracting a guarantee that the Reich would double its spending on the running costs within two years. Excavations began in October. One of the first structures to be built was the director's house, allowing Debye to reside on site.

Stark was not the only person unhappy about Debye's appointment. In June 1935 Felix Krueger, the new rector of the University of Leipzig, wrote to the American consul in that city protesting at the Rockefeller funding of an institute that was about to poach one of his most eminent professors, by whose departure, he said, the university would be 'seriously damaged'. Why build the institute in Berlin, he argued (to no avail), when Leipzig would provide a suitable site for it free of charge?

Debye harboured no sentimental attachment to Leipzig, however, and he was excited by the prospect of leading such a well-equipped and independent laboratory. When Tisdale visited him in October, he reported that the Dutchman was 'the only undepressed person I talked to in Germany'. Debye assured him that the state authorities would not get in the way: Rust, he said shrewdly, 'knows little, is pretty much worried, and . . . can, if properly handled, be influenced'. The following summer, Tisdale and Weaver stopped in Berlin en route to Holland to check on how the money was being spent, and Debye gave them a tour of the new institute, of which the exterior had by then been nearly completed. By 1937 scientific research had begun, even though the institute was not yet officially inaugurated.

Debye had two research priorities: experiments to investigate how substances behaved in large electric fields and at very low temperatures. In the summer of 1935 he visited the laboratories of Franz Simon at Oxford and Wander de Haas at Leiden, both of whom carried out low-temperature studies using the recently discovered technique of liquefying helium as a coolant. Debye was convinced that new kinds of physical behaviour could be discovered under such exotic conditions. At Leiden, Heike Kamerlingh Onnes had discovered in 1911 that metals can conduct electricity without any electrical resistance at very low temperature (the phenomenon of superconductivity), while the flow of liquid helium with no viscous impediment (superfluidity) was discovered at Cambridge University in 1937. Both are properties that result when quantum-mechanical principles begin to dominate the materials' behaviour, which happens only as the disruptive influence of heat is frozen out.

To provide the coolants for such experiments at the KWIP, a laboratory for making liquid air and liquid hydrogen was constructed in a building separated from the main block to minimize the damage of

The Kaiser Wilhelm Institute for Physics in Berlin-Dahlem, 1937. The tower housing the high-voltage equipment is on the left. Today this 'Lightning Tower' is a repository for the Archives of the Max-Planck-Gesellschaft.

a potential explosion. But it was the high-voltage apparatus that most commanded the eye. Electromagnets capable of generating 2.8 million volts, built by Siemens & Halske, were housed in a quasi-Romanesque tower twenty metres high at the western end of the main wing.

As well as Debye, Laue and a third professor, the experimentalist Hermann Schüler, the institute had six junior researchers. Debye brought his former assistants from Leipzig – Ludwig Bewilogua, Wolfgang Ramm and the Dutchman Willem van der Grinten – and also took on the chemists Friedrich Rogowski and Karl Wirtz, and Heisenberg's Leipzig student Carl von Weizsäcker. Weizsäcker had formed a close friendship with Heisenberg and would become something of a confidant. Over the next three years there were also many visiting guests from abroad.

The atmosphere was rather more liberal than at the universities. As Wirtz recalled, this was not so much because the directors were permissive but because they were too preoccupied with their own work:

> It soon became apparent that the individual younger employee was granted a relatively independent existence, and complete independence in his choice of topics. I found this to be both pleasant and difficult . . .

The high-voltage equipment at the KWIP.

These freedoms were in part motivated by the personalities of the former directors. Both Debye and von Laue were preoccupied with their own work. Both were theorists and their assistants were not needed for immediate help with their own work . . . I also think of that time as being very productive, because I was gradually forced to become independent.

Weizsäcker had similar recollections: 'Debye was a very liberal director. He didn't really give me a job, but told me that I should simply explore what interested me.'

Not everyone thought well of Debye's directorship. When Weaver visited Otto Warburg, still tenaciously maintaining leadership of the Institute for Cell Physiology in Dahlem, in 1938, he found him bitter and paranoid. Debye hadn't once called on him since he arrived, Warburg complained, only to contradict himself by saying that Debye had repeatedly tried to speak with him but that he had refused to enter into discussion. 'He said', Weaver reported back to his superiors,

> that the common notion that the government was irresponsible and wicked, while the professors were honest and idealistic is, in point of fact, exactly reversed, adding that the academicians are 'rotten to the bone'. He insisted that Kühn, Debye and Butenandt are interested only in things which they calculate will advance their own personal position.*

Warburg's assessment of Debye's self-interest echoes that of several others who knew him only a little (see page 172). He is, however, a decidedly unreliable witness: in his meeting with Warburg, Weaver 'got the impression of a man who . . . is very near the edge of mental instability', with a 'fairly well-developed persecution complex'. One can hardly be surprised at that: Warburg's 'non-Aryan' ancestry would have made his position precarious even without his courageous complaints to Rust about the disruptive effect on his assistants of compulsory Hitler Youth parades. And it was surely enough to make anyone feel paranoid that, just a week before Weaver's visit, Warburg had read his own obituary in *Nature*, having been mistaken for a namesake botanist who had emigrated from Berlin to Palestine.

How Warburg managed to keep his place at the KWI throughout the Third Reich is something of mystery, even if officially he was exempt from the Civil Service Laws because his institute was funded by the Rockefeller. Some say that he enjoyed good connections, others that the hypochondriac Hitler hoped he might find a cure for cancer. When Warburg was dismissed in 1941, the decision was successfully appealed by Viktor Brack, the chief of staff at the Reich Chancellery. This was the same Brack who helped to engineer the 'euthanasia' of more than 50,000 Jews, gypsies and mentally ill people – another example of the deep and perplexing contradictions in the Reich, and a reminder that we are unwise to seek tidy consistency in the motives of its protagonists. Brack's intervention undoubtedly saved Warburg

* Alfred Kühn was director of the KWI for Biology, Adolf Butenandt the director of the KWI for Biochemistry, both in Dahlem. Butenandt was a party member who obeyed the edict forbidding the acceptance of Nobel Prizes (see page 122), turning down the chemistry prize in 1939 for which he was nominated for his research on sex hormones. In 1949 he was happy to accept it retrospectively.

from the camps; but Brack told him that 'I did this not for you, nor for Germany, but for the world.'

Poison pen

As head of the prestigious new institute, Peter Debye evidently felt buoyant in 1936. While Planck seemed to be sinking further into despair, harassed by the 'Aryan' physicists who wanted him removed from leadership of the KWG, and while Heisenberg became increasingly isolated and demoralized at Leipzig, Debye had so far been able to shrug off most political interference and was riding the crest, apparently untouchable. At the end of the year his standing in German science was confirmed when he heard that he had been awarded the Nobel Prize in Chemistry.

It's not easy to describe in simple terms exactly what the award was given for. The announcement cited Debye's 'contributions to our knowledge of molecular structure through his investigations on dipole moments and on the diffraction of X-rays and electrons in gases'. In other words, this was one of those Nobels given for a body of work rather than a single discovery: through his studies of the interactions between matter, electricity and electromagnetic radiation Debye had helped to elucidate what atoms and molecules look like and how they behave. The prize confirmed Debye, once a prospective electrical engineer, as the world's most distinguished 'electrical engineer of molecules'.

However much they welcomed international prestige for their intellectuals, the Nazis came to rue this particular form of recognition. At first they were laudatory: two days after the announcement on 12 November, Debye received a telegram of congratulation from Bernhard Rust. It would have pleased Rust that Debye told reporters he could not have achieved it without German support: 'it would be fair to say: Germany and Holland have won the Nobel Prize for Chemistry in 1936 together'.

But on 23 November the 1935 Nobel Peace Prize was given retrospectively to the German writer Carl von Ossietzky, whose pacifism had consigned him to a concentration camp since 1933. Hitler considered this a blatant piece of politicking by the Norwegian Academy, and let it be known that the Reich held the Nobel organization in contempt.

In January 1937 the government announced that thenceforth no German might accept a Nobel Prize. (The Nobel Committee ignored the edict, for example by awarding Otto Hahn the chemistry prize in 1944.)

As a non-German, Debye was again exempt from this ruling. But nevertheless the German Foreign Ministry decided he should not attend the award ceremony. By the time they relayed that decision to Debye, he had anticipated as much and had cannily left already for Sweden, where he received his golden medal from the Swedish king Gustav V. So the German Embassy in Stockholm was forced into an unhappy compromise by cancelling any celebration in its premises.

Does this, as some have claimed, show Debye to be a man determined to undermine, defy and oppose the Nazis? Certainly it suggests that he had no interest in courting political favour, but that was clear enough already. It is equally possible to interpret Debye's actions over the Nobel as more evidence of his alleged egotism: he was not going to let politicking rob him of glory. Once again, we have only the facts, and they are such as to permit whatever interpretation one feels inclined to impose. There is nothing in Debye's response here that is inconsistent with the picture of a man simply determined to avoid political interference as far as he was able. As with Planck's insistence on commemorating Fritz Haber, this episode seems not so much an act of ideological defiance as a desire to do what one wishes. At any event, once garlanded, Debye let the matter drop: when he returned to Germany, he declined to offer future recommendations to the Nobel Committee.*

In the autumn of 1937 Debye was elected chairman of the German Physical Society (DPG). This became something of a poisoned chalice when, the following year, the Reich began to tighten its anti-Jewish laws, and the Ministry of Education announced the intention to put all scientific associations 'on the same footing' – in other words, to ensure that they no longer had any Jewish members. The DPG had long manoeuvred to maintain its independence. It was partly with this in mind that Laue's replacement as president in 1933 was the industrial physicist Karl Mey, who worked at Osram AG: a non-academic, it was

---

* Whatever Debye's behaviour after the award, it is very hard to credit Sybe Rispens' suggestion that Debye was keen to retain his Dutch citizenship because he considered it less likely that a German would be given the prize during the Nazi regime. As we shall see, his national status remained vitally important to Debye after his 1936 award.

thought, would be less susceptible to government pressure. Mey's successor after the statutory two-year term of office – and Debye's predecessor – was Jonathan Zenneck, director of the Deutsche Museum in Munich.

Autonomy in Nazi Germany was relative at best, and came only at the cost of making concessions. The DPG was slower than most scientific bodies in purging its Jewish members, but it obliged the regime in some other respects. It was formally monitored by the Reich Education Ministry, to which the society would dutifully submit its candidates for the annual Planck Medal. In 1938 these were the French physicist Louis de Broglie, who suggested that quantum particles such as electrons might show wavelike behaviour, and the Italian nuclear scientist Enrico Fermi, working in Chicago. Fermi was initially the DPG's preferred choice, but the society duly dropped him when the REM expressed concerns about his 'racial type': he had a Jewish wife.* Moreover, the DPG excluded 'non-Aryan' contributors from a special issue of the society's journal *Annalen der Physik* in spring 1938 to mark Planck's eightieth birthday. Debye objected at first to this censorship, but eventually acceded. Among those excluded was Debye's old friend from his days with Sommerfeld, Paul Ewald, who was officially 'a quarter-Jewish' and moreover was married to a Jew. Ewald, who left Germany to work in England towards the end of 1938, expressed dismay that Debye had permitted this ideological interference, to which Debye replied that it would have been impossible to reach any other decision.

Debye took the same line when similar restrictions were imposed on the celebrations of Sommerfeld's seventieth birthday at the end of the year. The Jewish physicist Ludwig Hopf at Aachen, who had studied with Sommerfeld at Munich, wrote to ask if he might attend, but was told that this would not be possible and that matters were out of his hands. 'I fear that these lines will not be pleasing to you', Debye wrote, 'but [I] consider it best for you to know how matters really stand.' Despite the absence of much solace or sympathy in this communication, Hopf was more understanding than Ewald: 'You obviously cannot swim against the tide', he graciously told Debye.

* The suggestion that selecting these non-German candidates was in itself an expression of defiance carries little weight, given the DPG's readiness to submit to vetting. However, there had also been controversy over the granting of the award to Schrödinger in 1937, since he had left Germany in 1933 in protest at the Nazi policies.

Although the DPG made no formal move to expel its Jewish members before 1938, their position became increasingly untenable as they lost academic posts. Many left of their own accord: around sixty-five had done so by the end of 1937. The Dutch Jew Samuel Goudsmit, an overseas member of the DPG in the United States, resigned in protest at that time, saying 'I am disappointed that the society has never as a whole protested about the sharp attacks on some of its most distinguished members.' This was a fair criticism. The position of the DPG on the dismissals had always been that it would try to help individual members so affected while making no complaint about the principle that had created their predicament. Again, the emphasis was on what was deemed to be both effective and proper; public protests were thought to be neither. The DPG's official proceedings made almost no mention of the expulsions, instead continuing to run meeting reports, obituaries and news about new members and business matters, as though nothing had changed. Even the emigration of such eminent figures as Franck and Born was not acknowledged.

Nonetheless, the DPG arguably managed to evade the total ideological alignment witnessed in other scientific bodies – it never truly had a Nazi-appointed president, for example, and was slower to purge its Jewish members than was the German Chemical Society. Such laxity can hardly be considered a significant act of resistance, but it did eventually provoke official disapproval. Ever since the REM had assumed oversight of the DPG, it had sent its representative Wilhelm Dames to the society's meetings. At the autumn joint meeting of the DPG and the German Society of Technical Physics (DTPG) in Baden-Baden in September 1938, Dames decided it was time to turn the screws. At this time the Nazis were preparing a new wave of anti-Semitic activity – the vicious Kristallnacht was just two months away – and Dames announced that all scientific societies were now to be 'invited' to comply with the 'implementation of the Aryan principle'.

Dames was disgusted by how the physicists were conducting themselves. Both the DPG and the DTPG, he wrote to State Secretary Otto Wacker on 3 October, 'have made only slight progress in their general National Socialist conduct'. At Baden-Baden, he said, the speech of Karl Mey, then president of the DTPG, 'conspicuously lacked National Socialist references'. Worse, in referring to the fiftieth anniversary of the discovery of electromagnetic waves Mey made reference to the half-Jewish Heinrich Hertz, despite having been 'thoroughly informed about the position and wishes of the involved ministries'. Then at the official banquet, Dames continued with indignation, Mey

> gave an impossible incoherent talk before 800 mathematicians and physicists about the most insignificant matters – he was doubtless already tipsy – at the end of which he did not propose the usual toast to the well-being of the Führer and the Reich, but drank the first glass to the societies instead . . . Dr Mey's blunder became particularly conspicuous when Prof. Esau [Abraham Esau of the University of Jena, director of physics for the Reich Research Council], with the best of intentions, made up for the toast to the Führer shortly after Dr Mey's speech.

All this, Dames insisted, offended several of the scientists present, who told him that 'it would be impossible for them to continue to participate in the societies if conditions there were not changed along with the Jew question regarding membership and contributions to the societies' publications'.

As a result, Dames presented Mey with an ultimatum for the DTPG and DPG (whose president Debye was not at the meeting). The two societies would merge into one, which would provide an excuse for having their articles revised 'to current requirements'. This would be done under the supervision of either the REM or the National Socialist League of German Technicians under Fritz Todt, an engineer and long-standing senior Nazi. (Mey decided that the former would be preferable.) The new regulations, said Dames,

> would also have to provide that only citizens of the Reich can apply for regular membership and foreigners would be admitted as special members, if necessary. Furthermore, it would also have to be pointed out that Jews may not be involved in the societies' journals, either as editors or as contributors. The acceptance of contributions from members of the Jewish race will be restricted to exceptions (and only if they are exceptionally valuable); and in the review section reviews of papers by Jews should also be avoided.

The exceptions in Dames' letter only made the proposed regulations more offensive, if that is possible, since they acknowledge that the alleged worthlessness of Jewish contributions to science is just a fiction to be abandoned when expedient.

Debye saw that these developments threatened to eclipse whatever autonomy the DPG could still claim. What is more, he was under pressure from members keen to parade their National Socialist credentials. Two of these, Herbert Stuart of the University in Berlin and Wilhelm Orthmann of the Industrial College of Berlin, organized a petition calling for the resignations of Jewish members. At this stage Dames' warning of impending changes was no more than verbal – there was never an explicit order from the authorities about expulsions, but only an 'anticipation of obedience'. But there could be no real doubt that the DPG was facing an ultimatum, and Debye seems to have concluded that more would be salvaged by being proactive than by waiting for an official command. So on 3 December he drafted a letter, which was discussed by the society's board and sent out on the 9th. The final version said:

> Under the compelling prevailing circumstances, the membership of German Reich Jews in the German Physical Society in the sense of the Nuremberg Laws can no longer be upheld. In agreement with the Board of Trustees I therefore summon all members who fall within this provision to inform me of their withdrawal from the society.
> Heil Hitler!
> Peter Debye

This letter has become the key exhibit for the retrospective prosecution of Peter Debye. Not only was he prepared to accede to this most overt of anti-Semitic measures, but he signed the letter with the Nazi salutation! For Sybe Rispens, who presented this letter in his 2006 book as though it was a revelation even though historians had long known of it, this was prima facie evidence of Debye's collaboration with the National Socialists. Others have called the DPG dismissal letter a 'turning point' in Debye's relationship with the state. But in truth it was no such thing. It has been rightly pointed out that one can deplore the letter without having to deplore Debye for writing it.

To deal with the slightest issue first: the letter simply could not have been sent without the 'incriminating' 'Heil Hitler'. As Mark Walker explains, 'In the mid-thirties all officials, including professors, were obliged to place that phrase at the end of their letters. Even Max von Laue, who was known as an anti-Nazi, used it in his letters.' Max Delbrück, who worked under Lise Meitner in Berlin before moving to the California Institute of Technology in 1937 on a Rockefeller fellowship to study genetics, made a wry remark on the issue apropos a letter written by Laue and Otto Warburg to Rust:

> The question was how would they sign it, with 'Heil Hitler' or not? The choice was either 'Heil Hitler' or the old conventional formula, '*Mit vorzüglicher Hochachtung*' (With our greatest respect). They discussed it for a while and finally Laue said, if he said 'with great respect' it would be just a big lie, so I assume they wrote 'Heil Hitler'.

It would be absurd to suggest that Debye approved of the letter.* His original draft had 'invited' or 'requested' the resignations rather than more forcefully 'summoning' them. On the other hand, even in the final version these withdrawals were still being requested, rather than the Jews being told that they were expelled. One could read that either way: a 'voluntary' resignation potentially allowed the Jewish members to preserve some dignity, but it could also be interpreted as a hypocritical attempt to appear less dictatorial. It seems clear, however, that Debye wanted the tone to be placatory, even apologetic. His eldest grandson Norwig Debye-Saxinger has said that Debye contacted the affected DPG members to convey his personal apologies. When the letter was sent, owing to a clerical error, to Lise Meitner, who had just fled the country to Denmark and Sweden and who was in any case an Austrian, Laue wrote to her on 19 December telling her to disregard it, saying that its content 'will not have surprised you'. He added that when he and Debye added the ambivalent 'under the prevailing compelling circumstances' – implying a reluctant acceptance of matters beyond their control – some of the National Socialist sympathizers in the DPG committee had threatened

---

* Debye's grandson Norwig Debye-Saxinger claims to have been told that after signing the letter, Debye later sighed to his wife 'We must move away!' But there is no indication that at this stage he had the slightest intention of doing that, not least because he was committed to the KWIP.

to expose them in *Das Schwarze Korps* – to which Debye had apparently replied 'I couldn't care less!' (The original German phrase is rather more delightful: '*Das ist mir Wurst*', literally 'That's sausage to me'.)

No, Debye would surely rather have not had to send the letter. The question is whether he should have let himself be forced to do so. If he had objected as president, he would sooner or later have been removed and almost certainly replaced by someone more politically acceptable; likewise if he had resigned. What, then, would such a gesture have achieved, except tighter political control of the society? Instead, when the REM finally told Debye officially in March 1939 that 'an immediate settlement of the Jewish question within the Physical Society would be very welcome here', the president was able to respond that this had already been addressed, and thereby to demonstrate that the society could take care of its own affairs without intervention.

But was the moral price worth paying for what was after all a sham independence? If Debye had indeed resigned, that would be regarded today as noble rather than self-destructive. But both Debye and Laue, who was also on the DPG board, saw things differently. The prevailing view was that, rather than indulge in gestures deemed self-centred and futile, one must sigh, act with regret, and go home telling oneself that there was really nothing else to be done. If we are to pass judgement, it must be on the moral failings of this capitulation to fate rather than with shrill accusations of anti-Semitism or collaboration.

Yet what is most troubling in the DPG's decision is the apparent absence of any moral self-examination at all. The society's treasurer Walter Schottky, a former student of Planck, had his eye primarily on the financial and international implications. He worried that the expulsions might provoke some foreign members to resign in protest, which would not only look bad for Germany but would also eliminate the 'quite considerable foreign currency receipts' of their subscriptions from the society's coffers.

It's not clear that Debye's letter allayed suspicions about the society's political soundness anyway. 'The DPG is still very backward and still clings tightly to their dear Jews', sniffed the Reich University Teachers League after Debye's letter was sent out. This Nazified group scoffed at the wording of the dismissal: 'It is in fact remarkable that only "because of circumstances beyond our control" can the membership of Jews no longer be maintained.' As though one should have to wait for such an exigency before dismissing them! As for Debye, Wilhelm Schütz of the University of Königsberg, a member of the DPG committee and a National Socialist, considered him disgracefully soft. In a letter to Herbert Stuart, Schütz wrote that 'the handling of the Jewish question by the DPG demonstrates that Debye lacks the necessary understanding for political questions, which is what we should have expected. At that time I tried and failed to get a clear position from the chairman and thereby come to a definitive solution of the problem.'

The letter did not in fact affect many of the DPG's members. Contrary to Rispens' suggestion that it resulted in the dismissal of a third of the society's membership, by the winter of 1938 rather few Jews remained. Estimates vary, but the numbers seem sure to have been very small; there are records of only six or seven members resigning in response to the letter by the start of January 1939.* Despite Schottky's concerns, no foreign members seem to have resigned in protest, although one must allow that they may not even have known about the affair. And those 'non-Aryans' who did tender their resignation seem largely to have considered that the DPG was faced with no alternative. One of them was the theoretical physicist Richard Gans, who later emigrated to Latin America. 'I can assure you', he wrote to a German colleague in 1953, 'that I've never felt bitter about my expulsion from the German Physical Society, because I knew that there was an act of *"force majeure"* against the will of the society.' When the exiled Jewish scientist Kasimir Fajans was interviewed during the FBI's investigations into Debye on his arrival in the United States in 1940 (see page 172), he admitted that he was disappointed with Debye for not having the 'moral stamina' to resign rather than sign the letter; but he nevertheless 'seemed to have a very high opinion

---

* Historian Klaus Hentschel suggested in 1996 that perhaps as many as 121 Jewish members were dismissed – about one tenth of the society's membership – but he later found that most of the names that disappeared from the DPG's membership records between 1938 and 1939 did so for other reasons, such as death or emigration. Eighty-four new members joined during that period, so the DPG did not change substantially in size. Moreover, politically motivated resignations had been happening since 1933, so it remains hard to establish which of those in 1938 were in direct response to Debye's letter.

of Debye', saying that he was 'interested only in science and not in politics'. Of course, the limited consequences and the forgiveness of the victims hardly mitigate the morality of the act itself. Rather, Gans' remarks say much about how the situation was regarded by all parties: almost as an act of nature, against which the individual was powerless.

In any event, Debye and his colleagues knew that the letter was merely completing a process that had already almost run its course. Debye himself remained determined to avoid 'political' matters as much as possible. He was always wary of meetings and speaking invitations that had a hidden agenda, for example withdrawing from an evening lecture in Danzig in 1939 when he heard that party members would be present.* There is little reason to believe that he cared to ingratiate himself to the Nazis.

Debye and the Jews

Perhaps the most damaging charge that the DPG letter has drawn against Debye is that it reflects an underlying anti-Semitism. In support of that idea, a letter from Debye to Sommerfeld in 1912 has been adduced in which he could be said to racially derogate the Austrian physicist Paul Ehrenfest:

> If you are thinking of getting Ehrenfest, I cannot refrain from expressing some reservations. A Jew, as he openly is, of the 'high priest' type can have an extremely harmful influence with his twisted Talmud logic. Many a bright, not completely ready idea, which would otherwise be expressed with bold courage, can only too easily be nipped in the bud that way.

---

* Debye was apparently troubled also that the topic of the meeting, low-temperature physics, would oblige him to mention the leading work in that field by Franz Simon in Oxford, which, since Simon was Jewish, would be politically compromising. Rather than submit to that censorship, he would rather not attend at all. While this could certainly be regarded as too ready an acceptance of Nazi prohibitions, it surely speaks also of a certain amount of integrity in this refusal to edit science. The conclusion of Martijn Eickhoff, author of a 2008 report on Debye commissioned by the Netherlands Institute for War Documentation (NIOD) in response to Sybe Rispens' book, that 'by exercising [this] self-censorship he indirectly demonstrated his loyalty to the Third Reich' typifies the slanted analysis in that report.

Others have tried to insinuate that by failing to curb the Jewish exclusions, Debye was somehow condoning them. At the KWIP, says Martijn Eickhoff (see footnote, p. 130), 'Debye engaged in an exclusive form of German science: physics were [sic] practised at the highest level and in principle there was no longer any place there for Jewish scientists with German nationality.'

If this is the sole evidence for Debye's 'anti-Semitism', it is scarcely worth bothering with. From today's perspective his remarks about Ehrenfest are hardly tactful – and misplaced anyway, for Ehrenfest was a physicist of the first rank – but at worst they exemplify the pervasive racial and cultural stereotyping of the early twentieth century. The same applies to Debye's remark to Sommerfeld, after Ehrenfest had secured an appointment at Leiden, partly on the recommendation of Einstein: 'I think that the racial issue played a part, even if perhaps at a more unconscious level.' There is here a hint of the common prejudice that the Jews look after their own, but such an immature remark by a young man is scant reason for a verdict of trenchant anti-Semitism.

No one who knew Debye, including Jewish friends and colleagues such as his student Heinrich Sack (see below), has recorded the slightest suspicion that he harboured antipathy towards Jews. Debye's son Peter insists that his father 'was not interested if [a] man was Jewish or not Jewish or whatever the situation was; he was interested if the man had good ideas' – a testimony that, if no more than what filial devotion might command, nonetheless rings true.

In his 2008 report on Debye commissioned by the NIOD, Martijn Eickhoff seemed determined to prove otherwise. Armed with a tiny collection of these and other injudicious but ambiguous comments from Debye's early career, he asserts that the absence of such comments in his later years shows Debye concluded that anti-Semitism would no longer work to his advantage. I suspect this may be the first ever suggestion that public displays of anti-Semitism could be bad for your prospects in Nazi Germany.

Eickhoff's flimsy insinuations not only defame Debye but also cloud the whole issue of how to think about the response of the German scientists to the oppression of the Jews. It is all too tempting to suppose that no one would have tolerated the anti-Semitic laws without protest unless they were secretly in favour of them, and likewise that no one would have helped a Jew unless they were an anti-Nazi activist. Mark Walker has lamented this insistence on simple formulae:

> Today, among . . . contemporary Germans, if a German saved one Jew or if he stood up for the ideas of one Jew once, then this man was not an anti-Semite. However, among several Jewish scientists, Jews of today, if a German once did not stand up for a Jew, or once helped persecute a Jew, or once helped persecute the ideas of a Jew, then this man is an anti-Semite. And there's really no chance for compromise there.

Such attitudes, Walker rightly implies, wholly misunderstand the situation. For one thing, there was no stigma to being an anti-Semite in Germany (or Austria, or indeed most of Europe) in the early part of the century, and the National Socialist regime removed any vestigial inhibitions on that score – indeed, they made anti-Jewish sentiment a social and professional virtue. So there is simply no reason why latent anti-Semitism need surface only in unguarded remarks made in private. The real problem was not that there were hordes of closet anti-Semites, but that those who were not tainted with that prejudice felt so little compulsion to deplore it. If you were not Jewish, then on the whole it was a matter that didn't concern you, even if you abhorred injustice and brutality. This is why Debye, like so many in Germany, was very ready to help his Jewish colleagues while making no public protest about the measures that had caused their difficulties. Heinrich Sack, for instance, was an able assistant to Debye in Zurich and Leipzig, and in 1933 Debye helped to arrange a position for him at Cornell University. He also assisted the Jewish chemist Hermann Salmang in finding a job with a Maastricht ceramics firm after being dismissed from a Kaiser Wilhelm Institute in 1935.

Far more notable was Debye's role in the departure from Germany of a particularly significant Jewish scientist, at no small hazard to himself. It was a loss that may have cost Germany dearly in wartime, but for which the rest of the world should be thankful. For shortly after she escaped, with Debye's help, to Copenhagen, Lise Meitner conceived of the theory of nuclear fission.

Escape from Berlin

Meitner was one of the few Jewish scientists who managed to retain an academic post until just before the war began. True, she was dismissed from her position at the University of Berlin in 1933, barred from speaking at scientific meetings, and all but erased from the official narrative of German nuclear physics during that time, so that her joint discoveries with Otto Hahn at the Kaiser Wilhelm Institute for Chemistry were attributed to him alone. Yet she was able to stay in active research at the institute until 1938.

When German troops entered Austria on 12 March that year to be greeted by adoring crowds, being an Austrian Jew in Berlin was no longer merely anomalous but perilous. Events in Vienna made it very apparent what the *Anschluss* implied: Jews there were turned out of their homes, many were brutally beaten and spat on in the streets, some were murdered. Nazi sympathizers at the KWIC no longer moderated their language – the fanatical Nazi chemist Kurt Hess, next to whom Meitner had had to live for several years, proclaimed that 'the Jewess endangers this institute'. It was an outrage to the likes of Hess that, while the institute for physical chemistry had long since purged its staff of non-Aryans after Haber's departure and aligned itself to the regime, the chemistry institute's director Hahn still permitted them to remain.

Although no one could accuse Hahn of having sympathy for National Socialism, his response to the crisis in 1938 does him no credit. He went to speak with the institute's sponsors, the Emil-Fischer-Gesellschaft, and returned on 20 March to tell Meitner that she must leave. He had been Meitner's closest colleague for twenty years; now he presented himself as little more than a courier bearing bad news. 'He has, in essence, thrown me out', Meitner recorded angrily in her diary. Hahn's wife felt shamed by the situation, which may have contributed to a nervous breakdown.

Meitner was arguably the best nuclear scientist in Germany, and neither the KWG's director general Ernst Telschow nor its president Carl Bosch wanted her to quit. Both they and she hoped that a way might be found for her to continue her research, which seemed to be on the verge of something important. Her friends abroad were deeply worried. Debye's former colleague Paul Scherrer wrote from Zurich inviting her to come and deliver a talk, and Bohr did likewise in Copenhagen; both were evidently offering escape routes from Germany. Yet still she hesitated, and weeks and months went by. By the time Meitner realized that emigration was the only realistic option and agreed to go to Copenhagen, where her favourite nephew Otto Frisch was working with Bohr, it was too late: she was refused a visa for Denmark.

Bohr, passing through Berlin on 6 June, was told by Debye that there was no great urgency about getting Meitner out of the country. He was mistaken: on the 14th Meitner learnt that not only was her resignation from the KWIC now expected, but that all technicians and academics were to be prohibited from leaving Germany. The Reich Interior Ministry wrote to Bosch on the 16th, saying that

> political objections exist to issuing a foreign passport to Prof. M[eitner]. It is considered undesirable that renowned Jews travel from Germany abroad to act as representatives of German science or even, using their name and experience, to act in accordance with their inherent attitude against Germany. The KWG could surely find a way for Prof. M to continue to remain in Germany following her resignation as well and, as the case may be, to also work privately in the interest of the society.

The note added that Himmler himself had confirmed this view – evidently Meitner's case was now known to him.

Debye wrote at once to Bohr in coded terms that never once mentioned Meitner but left no doubt about the meaning:

> When we last spoke, I assumed everything was quite all right, but in the meantime it has become clear to me that circumstances have substantially changed . . . I now believe it would be good if something could happen as soon as possible . . . I have taken the responsibility of writing all this myself, so that you can see that I too concur with the opinion of the concerned party.

Bohr passed the letter to the physicist Dirk Coster, an old friend of Meitner at the University of Groningen, who had discovered the element hafnium in 1923 while working in Copenhagen with Hungarian radiochemist Georg de Hevesy. Coster had been arranging emergency help for several refugee scientists coming to Holland from Germany, and had already written to Meitner in May to invite her abroad. He and his colleague Adriaan Fokker in Haarlem began seeking a position and funds for her, but with little success: most potential donors had already committed what financial resources they had. Coster and Fokker petitioned the Dutch government directly to permit Meitner's entry, and were granted permission when an unsalaried post was found for her at Leiden at the end of June.

But there was hardly any money to pay for her keep. With that in mind, Coster decided to go to Berlin to see for himself if Meitner's departure was absolutely necessary – not even he and Fokker yet grasped the real urgency of her situation. He wrote to Debye saying that he was coming to look for an 'assistant'. By coincidence, at the same time Meitner was offered a position in Stockholm alongside the Nobel laureate physicist Manne Siegbahn. She accepted, and Coster cancelled his trip, assuming that all was now in hand.

It wasn't. Meitner had planned to leave for Stockholm in August, but on 4 July Bosch told her that the plans to prevent scientists from leaving Germany were to be enforced imminently. It was now or never. Debye alerted Coster by letter on the 6th:

> The assistant we talked about, who had made what seemed like a firm decision, sought me out once again . . . He is now completely convinced (this has happened in the last few days) that he would rather go to Groningen, indeed that this is the only avenue open to him . . . I believe he is right and therefore I want to ask whether you can still do anything for him . . . If you come to Berlin may I ask you to be sure to stay with us, and (providing of course that the circumstances are still favour-able) if you were to come rather soon – as if you received an SOS – that would give my wife and me even greater pleasure.

It wasn't until 11 July that Coster received confirmation from officials in The Hague that Meitner would be admitted into the Netherlands. He set out at once for Berlin, where he stayed with the Debyes.

Only four people in Germany, aside from Meitner herself, knew of the plan to get her out: Debye, Hahn, Laue, and the science editor Paul Rosbaud, whose work for the KWG's journal *Naturwissenschaften* had brought him into close contact with most of the country's leading physical scientists. Coster had planned for Meitner to leave on the 13th; she spent the previous day working at the institute from early in the morning until 8 p.m., when she left to quickly pack her two small suitcases, assisted by a nervous Hahn. Rosbaud then drove the two of them to Hahn's house, where she spent the night. There Hahn made slight but poignant amends for his earlier failure to defend her by giving Meitner a diamond ring inherited from his mother, as an emergency fund.

After dark on the 13th, Rosbaud drove Meitner to the train station. There they met Coster, who boarded with her, and they travelled across the border without incident. It was nonetheless a deeply harrowing journey for Meitner, who at one point lost her nerve and begged Rosbaud to turn back as they headed for the station. 'At the Dutch border', she later recalled,

> I got the scare of my life when a Nazi military patrol of five men going through the coaches picked up my Austrian passport, which had expired long ago. I got so frightened, my heart almost stopped beating. I knew that the Nazis had just declared open season on Jews, that the hunt was on. For ten minutes I sat there and waited, ten minutes that seemed like so many hours. Then one of the Nazi officials returned and handed me back the passport without a word.

It was a narrower escape than even she recognized. Meitner's Nazi neighbour Kurt Hess had realized that something was afoot and sent a note to the authorities to alert them. Only delaying tactics by two sympathetic policemen prevented Meitner's arrest.

Once in Groningen, Coster sent a telegram to Hahn to say that the 'baby' had arrived. As the news spread, Wolfgang Pauli sent a characteristically witty note to Coster: 'You have made yourself as famous for the abduction of Lise Meitner as for [the discovery of] hafnium!'

Debye's defenders have argued that his actions in this case could not possibly be those of an anti-Semitic Nazi collaborator. And of course they could not. The political clampdown in 1938 meant that even Debye could no longer avoid the consequences of Nazi rule. But his courageous and humane intervention in Meitner's flight must still be seen within the broader context of life in the Third Reich. We should not imagine that Debye's assistance to Meitner, however praiseworthy, 'explains' anything: it was an act of human compassion towards a colleague, and does not in itself make Debye an anti-Nazi activist. Consider, for example, the case of Winifred Wagner, Richard Wagner's English daughter-in-law, who admired and befriended Hitler yet also saved several Jewish artists from the Gestapo.

In any event, Debye did not display much sensitivity in the matter. Even allowing for the caution that censorship of mail would recommend, his letter to Meitner in Sweden in November 1938 has a callow heartiness which seems to imply that her traumatic escape was just a brief distraction from the important business of doing science:

> I very much hope that by now you have found your feet in your new setting. That should not be difficult and with that everything has been settled. For as I know you, you will then automatically be completely happy because from that moment on you will be able to live entirely for science again.

On the other hand, Martijn Eickhoff's attempt to turn the Meitner incident against Debye is contrived and incoherent. It was, says Eickhoff, 'connected with a survival mechanism of ambiguity that Debye had developed and [was] primarily motivated by the desire to maintain the interests of his German science network; in the end [it] also rendered his own position secure'. How an action universally approved of by the few who knew about it supported a 'survival mechanism of ambiguity' is anyone's guess – Eickhoff seems to invoke the bizarre image of Debye contriving to help a Jew in one case and then banishing others from the DPG the next, purely to keep his options open and his colleagues guessing. Likewise, to suggest that Debye, Nobel laureate, head of the DPG and director of the KWIP, was somehow furthering his own interests by doing what other anti-Nazi scientists had been doing for several years is ludicrous. But the real failing of these accusations is to imagine that morality is a one-dimensional affair, a single axis along which our actions shift us between the poles of sainthood and depravity.

The spy

If we want to see what genuine opposition could look like in Nazi Germany, we should turn to Paul Rosbaud, one of the key orchestrators of Meitner's escape. Rosbaud was what we would now call a networker, intimately acquainted with most of the key physical scientists in Germany and abroad. His professional acumen in science communication and publishing led him after the war to set up the Oxford-based scientific and medical publishing house Pergamon Press with the later media magnate Robert Maxwell.

In National Socialist Germany, Rosbaud was not simply an anti-Nazi; he was a spy working for British intelligence, and his activities are somewhat inconvenient for those who argue that there was very little one could do genuinely to oppose Hitler's regime. Rosbaud opposed them in every way he could, at immense personal risk, and in exploits that seem plucked straight out of a *Boy's Own* post-war fiction. He joined the Nazi Party to gather information at high levels, he sometimes posed as a member of the German armed forces, and he supplied the Allies with important information on both the heavy-water operations for wartime nuclear research and the V-2 rocket work at Peenemünde. It is generally recognized that he was the informant code-named the Griffin by MI6, although, despite recent legal cases to force disclosure of official wartime secrets, this has never been officially confirmed.

Rosbaud's motive for aiding the Allies was simple: he despised Hitler's agenda. An Austrian from Graz, he studied science in Darmstadt and Berlin* before working for the mining, metallurgical and chemicals conglomerate Metallgesellschaft AG in Frankfurt. He became a scientific adviser for the Berlin-based metallurgical magazine *Metallwirtschaft*, in which capacity Rosbaud began to travel widely to visit scientists in Oxford and Cambridge, Copenhagen, Oslo and elsewhere. He got to know Einstein, Bohr, Rutherford, Hahn and Meitner,

---

* Rosbaud's biographer, the Manhattan Project scientist Arnold Kramish, says that after Darmstadt Rosbaud was granted a fellowship to study X-ray cinematography – a now obsolete discipline concerned with the use of X-ray imaging in medicine – at the 'Kaiser Wilhelm Institute in Dahlem'. It isn't clear which institute Kramish had in mind – the KWIP was not at that stage (the 1920s) functioning as a research centre. The topic of research seems odd for a physical scientist; others have assumed he was in fact working on X-ray diffraction.

and served as an adviser to various scientific organizations in Europe. His horizons were broad indeed: his brother Hans was a leading conductor in Germany and a friend of Paul Hindemith, and Rosbaud enjoyed the lively, permissive milieu of Weimar Berlin, befriending artists such as the Bauhaus director Walter Gropius.

Some time in the early 1930s Rosbaud met the English intelligence agent Francis Edward Foley, who was working under cover at the British legation in Berlin as a passport control officer. It seems likely that their acquaintance began after the Nazis came to power, when Rosbaud began helping Jews to leave Germany and encountered Foley doing the same thing. Rosbaud never knew if he was 'officially' Aryan himself – he was illegitimate and had no knowledge of his father. This meant he was unable to provide evidence of his Aryan heritage as required in 1933. But he knew how to exploit the inefficiencies, loopholes and laziness of the Nazi bureaucracy, and simply enlisted an old family friend in Graz to masquerade as his father.

Rosbaud started to pass potentially useful information to Foley on an informal basis. Now an adviser to the publisher Springer Verlag, which produced the multidisciplinary *Naturwissenschaften*, Rosbaud was well placed to gather details of military-oriented research in Germany. *Naturwissenschaften* was edited by the Jew Arnold Berliner until Springer bowed to government pressure and removed him in 1935. To the disgust of the National Socialists, who arranged a boycott of the journal, it continued to accept articles by Jewish authors.

By April 1938 it became clear that neither Rosbaud's wife Hilde nor their daughter Angela could safely stay in the country. With Foley's help, Hilde obtained a visa for England, where she was soon joined by her daughter. Rosbaud could have left too, but elected to remain and fight the Nazis. Besides, the arrangement suited him, for it meant that his long-term lover could move into his house; he cared for his family, but his were the ways of a Weimar libertine. In 1939 he colluded with Foley to secure the 'denial' of his own application for an English visa, establishing a convincing cover for his informant activities. Around 1940 his link with the British Secret Service was made official: he subsequently reported to Eric Welsh of MI6, who oversaw much of the intelligence-gathering on German science.

It's not clear what Rosbaud told the Allies during the war. Arnold Kramish alleged that he was the author of the anonymous Oslo Report sent to the British legation in late 1939, an important document that described a range of German 'secret weapons' and military strategies. However, the British officer in charge of scientific intelligence, Reginald Victor Jones, revealed in his memoirs published in 1989 that the report was the work of the industrial physicist Hans Ferdinand Mayer. Rosbaud did take part in the 'day of wisdom' visit to Peenemünde orchestrated (perhaps with a distinct lack of wisdom) by Wernher von Braun in 1941, allowing him to supply Welsh with a report on the rocket programme. In 1942 he travelled to Oslo in military uniform (probably of the Luftwaffe) to pass on information about German nuclear research to the Norwegian resistance, from where it could reach Welsh. It was partly through Rosbaud's efforts that the Allies kept track of the German efforts to harness nuclear energy using heavy water made at a Norwegian hydroelectric plant, which was consequently the focus of attacks by the resistance and British bombers.

How Rosbaud survived the war without being discovered seems to have been a mystery even to him. 'The last years have not passed without leaving marks upon me', he wrote to his brother Hans in 1946. 'There were too many in the underground who could not be saved and, at the end, only I slipped through by a hair's breadth. My hatred of the Nazis has not diminished.' Bound by British secrecy law, and in any case not one for self-glorification, he said nothing subsequently about his clandestine wartime activities.

That Rosbaud knew Debye and maintained cordial relations with him after the war is unremarkable in itself, for Rosbaud knew everyone. Nonetheless, this friendship has been advanced by chemist Jurrie Reiding in defence of Debye's good character. Reiding has even claimed that Debye may have given Rosbaud information about German military research in the late 1930s. 'Debye moved, as a prominent scientist and science manager, in higher Nazi circles', Reiding writes. 'He was on the board of the German Academy for Aviation Research and met Goering personally. Debye must have had thorough knowledge of German war technology . . . Therefore, the hypothesis that Debye was a secret informant for Rosbaud does not appear too bold.'

Neither, sadly, does it appear to be anywhere supported by hard evidence – Reiding can adduce only some ambiguous statements in a letter from Rosbaud to Debye after the latter had left Germany for the United States in 1940. Besides, although there is no doubt that

Debye disliked the Nazis, friendship with Rosbaud is no gauge of political persuasion: for example, he also seems to have felt genuine regard for the geologist Friedrich Drescher-Kaden, who was an ardent Nazi.

If we wish to find heroes in the tale that this book relates, Rosbaud comes as close as anyone. But it is of doubtful value to demand why there were not more like him among the German physicists. Not only is this degree of courage and resourcefulness exceptional, but the idea of actually aiding 'the enemy', rather than merely trying to moderate the excesses of the German leaders, would have been anathema to most Germans, insistent as they were on the false distinction between loyalty to the fatherland and loyalty to the government. It's more instructive to recognize that there was no ambiguity about Rosbaud – we do not need to piece together his attitude towards the Nazis from hints, stray comments, ambivalent actions. This is what true active opposition and moral responsibility looked like. Rosbaud did not deplore the weaker responses in other less resolute individuals but, as we will see, he was a rather astute judge of character and not easily deceived by retrospective self-justification. He saw what went on, he spoke his mind, and he offers one of the most reliable moral compasses through the maze.

8 'I have seen my death!'

When Max Planck promised Reich Minister Joseph Goebbels that the new Kaiser Wilhelm Institute for Physics would explore new areas in 'the field of atomic physics', leading to 'revolutionary innovations in public life', he was alluding to possibilities in nuclear physics that he and his colleagues had only just begun to glimpse. They had discovered that the atomic nucleus harbours unimagined energy, and they suspected that a way might be found to unlock and harness it.

Even allowing for the compression of time that the backward glance imposes, the path from the discovery of atomic structure to the obliteration of Hiroshima and Nagasaki seems frighteningly rapid. It happened within a generation: Marie Curie, who was there at the start, could reasonably have been expected to witness the devastation of the Japanese cities, had she not succumbed eleven years earlier to a particularly lethal form of anaemia induced by exposure to nuclear radiation in her research.

Until the Second World War began this was an international story, and as much collaborative as competitive. Discoveries made in Cambridge or Paris would be discussed within days in Berlin or Berkeley. In 1939 that ceased, and important discoveries in the nature and uses of nuclear power were regularly withheld from the science journals. It became a guessing game what one's foreign peers were up to, what they knew, what they were trying to make. But everyone knew what the stakes were.

Invisible rays

Like quantum theory, nuclear physics began with a puzzle about radiation; indeed several puzzles, as one thing led to another. In the *fin de siècle*, invisible and intangible rays seemed to be everywhere. First there were cathode rays, studied by Lenard at Heidelberg and explained by J. J. Thomson in Cambridge. These led to the discovery of X-rays by Wilhelm Röntgen at the University of Würzburg, who noted that when cathode rays struck the glass wall of a cathode-ray tube, it was not just the glass that fluoresced – the influence extended further. Thomson and Lenard had already noticed that; in 1894 Thomson observed that glass tubing would glow even a few feet away from the discharge tube. That couldn't be the work of the cathode rays themselves, which were stopped by the glass wall. Nor was it an effect of the fluorescent light from the glass: in 1895 Röntgen shielded the cathode-ray tube with thick black paper, yet still a phosphor screen beyond it glowed. If, however, he held up his hand between the tube and the screen, the phosphor image revealed the shadow of his bones. These mysterious penetrating rays, which Röntgen called X-rays, were apparently being *produced* when glass was stimulated into fluorescence by cathode rays. They could darken photographic plates, capturing the ghostly imprints permanently. Shown the X-ray of her skeletal hand, wedding ring and all, Röntgen's wife exclaimed 'I have seen my death!'

X-rays brought Röntgen a Nobel in 1901. Henri Becquerel in Paris heard about them in January 1896, and as an expert on fluorescence he wondered whether they might be emitted by fluorescing substances other than glass. One such was a salt of uranium, uranium potassium sulphate, which glowed after being exposed to sunlight.* Becquerel proposed to detect X-rays coming from the fluorescing uranium salt by their effect on photographic emulsion. He wrapped a photographic plate in black paper, masked a part of it with copper foil cut into a cross (which, he reasoned, should stop X-rays), scattered the salt on top and left it exposed to the sun to activate the effect. And indeed he found that when the plate was developed, the image of the cross was imprinted upon it.

But no X-rays had done this, as Becquerel subsequently realized in a famous stroke of serendipity. Having tried to repeat the experiment

* This glow is technically not fluorescence but phosphorescence. Fluorescent materials glow only while they are being irradiated, for example with light or X-rays, while phosphorescent substances can capture and store energy, continuing to glow even in the dark.

on an overcast February day, he put the plate, with the copper mask and uranium salt still on top, into a drawer for several days before deciding to develop it anyway, expecting to see at best only a weak imprint. Instead, the image of the copper cross was as clear as before. Sunlight wasn't needed to activate the process, then; unlike Röntgen's glass, Becquerel's salt was giving off radiation spontaneously, without any apparent stimulation. This was something new.

Becquerel's 'uranic rays' weren't perceived as having the allure of X-rays, and the finding was not widely pursued. It did, however, spark the interest of Marie Curie, a Polish woman (*née* Maria Sklodowska) who had come to Paris to study science and mathematics at the Sorbonne. Marie married Pierre Curie, an instructor at the School of Chemistry and Physics, in 1895, and the birth of their daughter Irène in September 1897 prevented her from starting her doctorate until early the following year. She figured that Becquerel's rays, which she named 'radioactivity', would make a good subject, 'because the question was entirely new and nothing yet had been written upon it'.

Becquerel had noted that the uranic rays made air electrically conducting: they were capable of ejecting electrons from atoms, leaving the atoms charged (ionized). Pierre Curie had some years earlier invented an instrument for measuring electrical charge very accurately, called an electrometer, and the Curies now used this as a means of quantifying the activity of uranium. At first they used relatively pure uranium salts for their studies, but when Marie tested raw uranium ore (pitchblende, mined in Saxony), she found to her surprise that it was even more radioactive. She concluded that the ore must contain a second radioactive element with even greater activity than uranium itself. She had already discovered that uranium was not unique: the 'uranic rays' were also detected from the rare element thorium. But now she set about chemically analysing pitchblende to identify the source of the extra radioactivity. This meant using chemical separation techniques to extract the minor impurities that contained the additional radioactivity, repeating the process again and again on several tons of laboriously crushed, dirty brown pitchblende to collect as much of the trace material as possible. As they gradually made the solution of this impurity more concentrated, the Curies saw its activity increase. In July 1898 they presented their findings in a paper read by Becquerel to the Institut de France. Here they claimed to have identified a new radioactive element, which they named after Marie's homeland: polonium.

There was a second radioactive impurity in pitchblende too, which showed different chemical behaviour and so could be separated from polonium. It was even more active – around a million times more so than uranium, enough for concentrated solutions to glow spontaneously as the 'uranic rays' excited fluorescence in the water. That accounted for the name that Pierre recorded in his lab book in December: radium.

Then radioactivity seemed to be everywhere. Electrometer measurements indicated that air itself was continually being ionized, and it was assumed that this was due to radiation streaming from natural radioactive elements in the ground or the air, such as the recently discovered inert gas radon (which is found in some types of rock). However, in 1912 the Austrian physicist Victor Hess in Vienna sent electrometers up in a balloon and discovered that the rate of ionization increased rather than decreased as the instruments were carried higher. The ionizing rays were coming from space: they were christened 'cosmic rays'. Hess won the Nobel Prize in Physics in 1936, the same year that Peter Debye was rewarded in chemistry. Since Hess was a Jew, the National Socialists were unimpressed. Two years later he was arrested in Graz after the *Anschluss* because of his refusal to accept Nazi rule; he escaped and emigrated to the United States. Cosmic rays were of great interest to nuclear physicists, in part because they offered a source of very-high-energy particles as projectiles for nuclear-transmutation experiments, exceeding the energies that could be reached in particle accelerators. Werner Heisenberg devoted much attention to the subject from the late 1930s.

Atomic energy

What was radioactivity? It was unnerving, to say the least. X-rays were produced only in response to some energetic stimulation, such as bombardment with cathode rays. But uranium just went right on discharging its energetic rays, day after day, even in the dark and cold, no matter how the element itself might be transformed into different physical or chemical states. Unlike chemical energy, the energy of radioactivity seemed inexhaustible. Two German physics teachers,

Hans Geitel and Julius Elster, took radioactive samples to the bottom of an 850-foot mineshaft in the Harz mountains of Saxony to test the hypothesis that they might be absorbing and then re-radiating rays that permeated all of space: they figured that a blanket of so much rock would attenuate any such influence. But it made no difference: the samples were as active as ever. The energy, they concluded, must be coming from the atom itself. It was *atomic energy* – and to understand it, one needed to understand the atom.

At the turn of the century, no one knew more about the atom than Ernest Rutherford, who arrived from New Zealand at Cambridge's Cavendish Laboratory in the year X-rays were discovered. There he began experimenting on Becquerel's uranic rays, and in 1899, shortly after moving to McGill University in Montreal as professor of physics, he reported that the rays are of two sorts: one, which he called alpha rays, are stopped by aluminium foil, whereas the second sort, called beta rays, are more penetrating. Rutherford later deduced that both of these varieties are actually particles, not rays. The beta particles are equivalent to cathode rays, that is, electrons. And in 1908, after leaving McGill to work at the University of Manchester, Rutherford showed in a beautiful experiment that alpha particles are the positively charged nuclei of helium atoms.

How can helium nuclei come *out of other atoms*? Rutherford had already realized that this involved the transmutation of one element into another. Experiments on thorium radioactivity were complicated by the fact that they gave inconsistent results unless the thorium was enclosed in a metal box. In 1899 Rutherford realized that this was because a gas was escaping from thorium and would, unless prevented, carry some of the radioactivity away. What was this 'thorium emanation'? That was a chemical question, and so Rutherford enlisted the help of a McGill chemist, an Englishman named Frederick Soddy. But Soddy's answer seemed scarcely credible. The emanation didn't undergo any chemical reactions at all, for it seemed to be nothing other than the inert gas argon discovered six years previously. 'Rutherford,' Soddy remembered stammering to his collaborator, 'this is transmutation – the thorium is disintegrating.' Rutherford boomed back that 'they'll have our heads off as alchemists!'

The thorium emanation wasn't in fact argon, but the much heavier inert gas radon, into which thorium is transmuted by radioactive decay.

Nonetheless, the principle of transmutation remained. Rutherford and Soddy realized it meant that radioactive emissions remove a part of the fabric of the atom, changing its chemical identity. As Rutherford's student Henry Moseley showed in 1913, the 'atomic number' that defines the place of every chemical element in the sequence of the periodic table – from 1 for hydrogen to 92 for uranium – is not just an arbitrary label but quantifies the number of positive charges (relative to the hydrogen nucleus) in the atom's nucleus. If a radioactive atom emits an alpha particle, this carries off two of those charges (the atomic number of helium is 2), and so reduces the atomic number by two, transmuting the element to that two places to the left in the periodic table. In this way, thorium, with atomic number 90, is transmuted to radium, with atomic number 88.* In other words, a radioactive atom 'decays' into another element as it emits radiation. (The emission of beta particles also induces transmutation, but that is a more complex matter which puzzled Rutherford and Marie Curie for years.) Some decays are so rapid as to be almost instantaneous, others are geologically slow. The rate is measured by the so-called half-life: the time taken for half of the atoms in a sample of a radioactive element to decay. This is always the same regardless of how much material you have. The half-life of uranium is about 4.5 billion years,† about the same as the age of the earth; that of thorium is just twenty-two minutes.

By radiating energetic particles (or high-energy photons called gamma rays in a third form of nuclear decay), radioactive atoms shed energy. In 1903 Rutherford and Soddy estimated the astonishing quantity of energy locked up in the atom, in comparison to which the energy released by any chemical process, such as the detonation of an explosive, is puny – at least 20,000 times less. Suppose, Rutherford mused, one could find a 'detonator' to expel all this atomic energy at once: then 'some fool in a laboratory might blow up the universe unawares'.

---

* This is just the first step in a whole series of decays, which proceeds via radon (element 86). The radon, being a volatile gas, will escape, but the disintegration of nuclei continues until it reaches the stable, non-radioactive element lead (element 82).
† As we will see shortly, there are in fact different types of uranium with different half-lives; this is the half-life of the most abundant natural form.

Soddy returned to England, and in 1904 he delivered this sobering thought in a lecture. If this energy 'could be tapped and controlled', he said,

> what an agent it would be in shaping the world's destiny! The man who put his hand on the lever by which a parsimonious nature regulates so jealously the output of this store of energy would possess a weapon by which he could destroy the earth if he chose.

Soddy's audience on that occasion was surely captivated by the thought: it was the Corps of Royal Engineers of the British Army.

The Curies had come to the same disturbing conclusion. In 1903 they and Becquerel were awarded the Nobel Prize in Physics for their work on radioactivity, but had been too busy and too ill to travel to Sweden for the ceremony (the kind of event that Pierre, in any case, loathed).* Marie gave birth to their second daughter Eve at the end of 1904, and so it was not until June 1905 that they finally came to Stockholm, where Pierre delivered the traditional Nobel lecture. 'It can even be thought', he said,

> that radium could become very dangerous in criminal hands, and here the question can be raised whether mankind benefits from knowing the secrets of Nature, whether it is ready to profit from it or whether this knowledge will not be harmful for it.

Pierre was optimistic about that, but one would not have guessed it from his demeanour. He was prematurely aged, constantly tired and often depressed. It wasn't clear why; he attributed his aches and pains to rheumatism. Marie too struggled with lethargy – she is, Pierre wrote to a correspondent in 1903, 'always tired without being exactly ill'. Her second pregnancy had ended that summer in the premature birth of a baby who died soon after. The Curies noted that their experiments with radium could leave them with lesions and reddened skin on their fingers; Rutherford noted that Pierre's hands were 'in a

---

* The prize was initially going to be given just to Pierre and Becquerel; only a late intervention from the Swedish mathematician Magnus Mittag-Leffler, a progressive advocate of women's rights, ensured that Marie was included. Her husband might have otherwise declined.

very inflamed and painful state due to exposure to radium rays'. Yet for Pierre these ailments never took their likely course, as a year after his triumphant lecture in Stockholm he was killed in a road accident in Paris. Had he lived, he would surely have shared in the second Nobel awarded to Marie alone in 1911, this time in chemistry, for the discovery of radium and polonium.

The many faces of atoms

If you wanted to know about radiochemistry and the structure of the atom, Rutherford was the man to ask. The 26-year-old Otto Hahn went to Montreal in 1905 to work with him, and at Manchester Rutherford took on the German physicist Hans Geiger. Rutherford and Geiger developed a device that revealed the passage of an alpha particle by its ionization of air, triggering an electrical discharge between two electrodes with an audible click – this was the predecessor of Geiger's famous counter. Geiger collaborated on the demonstration that alpha particles are ionized helium atoms in 1908, and the following year he, Rutherford and student Ernest Marsden conducted Rutherford's most celebrated experiment in which he deduced that atoms have small, dense nuclei. They shot a stream of alpha particles at thin gold foil, expecting them to pass through with some slight deflection. Mostly that's what happened; but to his perplexity, Marsden found that a very few of the particles bounced right back. 'It was quite the most incredible event that has ever happened to me in my life', Rutherford wrote later. 'It was almost as incredible as if you fired a 15-inch shell at a piece of tissue paper and it came back and hit you.' He concluded that this was only possible if atoms were not like little spheres – the 'plum puddings' proposed by J. J. Thomson, with electrons embedded in a suet of positive charge – but instead had most of the mass concentrated in a tiny, very dense and electrically charged centre, surrounded by a diffuse cloud of particles of the opposite charge. Mostly the atom was quite empty. Rutherford finally decided that the nucleus must have a positive charge, and the diffuse cloud contained electrons. This, he discovered, was much like the 'Saturnian' atom proposed in 1903 by the Japanese physicist Hantaro Nagaoka, in which rings of electrons orbited a positive (but by no means diminutive) core. As we saw earlier, this classical picture of a 'planetary' atom was soon reformulated in terms of quantum theory by Niels Bohr, who joined Rutherford at Manchester in 1912 'to get some experience in radioactive work'.

To understand the source of radioactive energy, one needed to understand the nucleus, which in Bohr's quantum atom could be regarded as just a ball of positive charge. What was actually in there? In 1815 the chemist William Prout had proposed that the hydrogen nucleus (later recognized as a single positively charged particle called a proton) was the fundamental building block of all the other atoms. If so, then the masses of all atoms should be simple multiples of the mass of hydrogen. This was roughly true in general, but not exactly, and sometimes hardly at all. Carbon and oxygen atoms, for example, are about twelve and sixteen times the mass of hydrogen – but chlorine atoms seemed to be about 35.5 times that mass. So Prout's law seemed highly inexact, and was largely ignored.

In 1919 Francis Aston at the Cavendish Laboratory, where Rutherford had recently been appointed director, devised an instrument for measuring atomic masses very accurately, which became known as a mass spectrometer. Aston was able to separate out atoms of different mass by removing electrons to make them electrically charged (ions), using an electric field to accelerate them down a channel evacuated of air, and then using a second field to bend their trajectories. If they have the same charge, then ions of different mass are deflected to different degrees and can be collected separately. Aston found that a pure element placed in the spectrometer could be separated into fractions with several different masses, each of them pretty much an exact multiple of the mass of hydrogen. Sulphur atoms, for example, could have masses of 32, 33 and 34, while chlorine atoms have masses of 35 and 37; their weighted average in a mixture of many billions of atoms resulted in the apparent 'fractional masses' that had seemed to flout Prout's law.

The idea that atoms of an element could differ in mass wasn't entirely new. Before the First World War interrupted Aston's research by billeting him for military work with the Royal Air Force, he acted as an assistant to J. J. Thomson at the Cavendish and developed physical techniques to separate atoms or molecules of different mass, based on the fact that in gases such particles diffuse at different rates – lighter particles move more quickly, like children darting nimbly through a crowd. While investigating the inert gas neon, Aston found tantalizing evidence for *two* forms of the element with slightly different atomic masses of 20 and 22.

Meanwhile, other researchers discovered that uranium decays into radium via an intermediate element, hitherto apparently unknown, which became called ionium. But ionium turned out to have chemical properties indistinguishable from thorium. In 1912 Frederick Soddy, now at Glasgow University, showed that the two substances emit light of the same frequencies – these frequencies were generally regarded as a fingerprint of chemical identity. Soddy therefore proposed that ionium was indeed a form of thorium, chemically the same but otherwise somehow distinct. Aston's studies of neon strengthened Soddy's conviction, and he proposed to call these different forms of the same element 'isotopes', meaning 'same shape'. Most chemists didn't like that idea – an element was supposed to be fundamental, not to come in different flavours – but Bohr, then with Rutherford in Manchester, endorsed it.

Aston's mass spectrometer now confirmed the reality of isotopes: those of neon had precisely the masses 20 and 22, as he had suspected. In 1922 Aston and Soddy were jointly awarded the Nobel Prize in Chemistry for their discovery – even though it was still unclear *why* isotopes differed in mass.

But Aston's measurements were precise enough to reveal something more: the atomic masses were *still* not quite exact multiples of that of hydrogen. They were always slightly less: there was a 'mass defect'. Aston and his peers realized that in this tiny deficit resided the immense power of the nucleus. As protons come together and *fuse* to form a nucleus, a little bit of their mass is converted to energy, in accord with Einstein's equivalence of mass and energy $E = mc^2$. The release of this energy is what makes the nucleus stable: it is called the binding energy. As Einstein's iconic equation implies, the energy equivalent of mass is enormous, being multiplied by the speed of light squared. From the minuscule mass defect of atomic nuclei, Aston could calculate how much energy was released by the fusion of hydrogen nuclei to form heavier elements. It was a phenomenal amount. 'To change the hydrogen in a glass of water into helium', he wrote, 'would release enough energy to drive the *Queen Mary* across the Atlantic and back at full speed.' If that source can be tapped, 'the human race will have at its command powers beyond the dreams of science fiction'. And again, those numbers were fearful: 'We can only hope that [man] will not use it exclusively in blowing up his next-door neighbour.'

The nuclear mass defect increases as atoms get heavier, but only so far as masses of about 60 (around the atomic mass of iron). After that, the mass defect gets steadily smaller. This implies that nuclei lighter than iron can become more stable – can acquire more binding energy per nuclear particle – by binding more protons, while those heavier than iron can do so by shedding protons. In other words, there are two ways for atoms to decay to more stable forms: by fusing or by disintegrating. Here, then, was an explanation for radioactivity. Very heavy atoms, such as uranium, have relatively less binding energy: they are liable to shed nuclear particles (alpha particles) to lower their mass and become more stable. In much the same way that energetic chemical compounds such as nitroglycerine are liable to undergo reactions to form more stable, lower-energy substances, so radioactive elements undergo nuclear reactions to the same end. By the same token, light elements are apt to engage in nuclear fusion, releasing vast quantities of energy in the process. In the 1920s it was recognized that this could answer the long-standing puzzle of how the sun and other stars could maintain such a prodigious output of energy for so long: they were powered by nuclear fusion, particularly of their main constituent, hydrogen.

Nuclear alchemy

Chemists could access, control and manipulate the energy release of chemical processes, but the machinations and transmutations of nuclei seemed to be hidden away out of reach, in the unthinkably tiny volume inferred by Rutherford. And yet, there was a way to get inside. Rutherford, Geiger and Marsden had watched alpha particles being deflected as they were repelled by the positive electrical charge of other nuclei. Yet if the alpha particles had enough energy, they could push through this repulsive barrier and enter the nucleus, which should itself be a favourable outcome in the case of lighter nuclei because of the binding energy it would set free. For an alpha particle, you might say, it is hard to get inside a (light) nucleus but worth the effort. Capture of alpha particles by nuclei raised the prospect of harnessing radioactivity to effect transmutation artificially: to convert one element into another at will.

That is what Rutherford achieved at the Cavendish in 1919. He showed that alpha particles emitted by polonium and passing through nitrogen gas could transmute some nitrogen atoms to oxygen. Alpha particles, being helium nuclei, have two protons, while oxygen has only one more proton than nitrogen. So the alpha particle is first subsumed by the nitrogen nucleus, and one proton is then spat back out alone.

This brand of nuclear alchemy had its limits, because as a nucleus gets more massive and more highly charged, the electrostatic barrier impeding an alpha particle's entry and fusion gets higher, and greater energy is needed to overcome it. The answer was to speed up the projectiles, granting them more energy. In 1929 the American physicist Ernest Lawrence at the University of California at Berkeley began to use electrodes charged to high voltages to accelerate particles – be they alpha particles or lone protons – so that they could push more forcefully at the electrical barriers of nuclei. Particle accelerators like Lawrence's could be used to induce transmutation in heavy elements, and were thereby tools for making new elements.

There was still something missing from the picture, and in retrospect it seems surprising that the nuclear physicists got so far without it. The proton or hydrogen nucleus could not be the sole constituent of the nucleus, for the mass of nuclei typically exceeded that of their protons. An alpha particle or helium nucleus, say, has twice the proton's charge but four times its mass. What produced this extra mass, typically about equal to the total mass of the nucleus' protons?

The answer is a second nuclear constituent, the neutron: a particle with no charge but a mass essentially equal to the proton's. Rutherford speculated in 1920 that such an entity – an 'atom' of mass 1, but without nuclear charge, as he expressed it – might exist. If so, what a boon it would be. Feeling no electrical repulsion from protons, it could enter other nuclei easily and thus be highly useful as a probe of the interior. But Rutherford did not regard this putative neutron as a fundamental particle; rather, it was a composite, a close association of a proton and an electron. His student James Chadwick was taken with the idea, and set out to find the neutron.

It was, however, first sighted in Germany. In the late 1920s, Walther Bothe and Herbert Becker in Heidelberg were using alpha particles from polonium to bombard light elements such as lithium and beryllium. Heavier elements such as boron and magnesium are disintegrated by this treatment, spitting out protons just as in Rutherford's experiments on nitrogen. But although beryllium was not disintegrated, nevertheless it emitted a kind of 'radiation' with intense penetrating power. Chadwick was intrigued, and so were Irène Curie, first daughter of Marie and Pierre, and her husband Frédéric Joliot in Paris. The French scientists found that Bothe's 'rays' could knock protons out of hydrogen-rich substances such as water and paraffin wax. They thought that the rays were gamma rays, although others doubted that gamma rays would have enough energy to eject protons. This was their hypothesis in a paper that the Joliot-Curies presented to the French Academy of Sciences in January 1932. Chadwick saw it the following month, and didn't believe it. He was convinced that in these findings 'there was something quite new as well as strange', and later admitted that 'my thoughts were on the neutron'. He applied himself to his own experiments, showing in short order that all could be explained on the assumption that Bothe's rays were indeed composed of particles like those proposed by Rutherford twelve years earlier, with zero charge and the same mass as the proton.

The neutron sets a great deal in order. It accounts for the rest of the nuclear mass: a nucleus consists of a number of protons (which determines the atomic number and chemical identity of the element) combined with a comparable number of neutrons. The carbon atom, for example, has six of each, and thus atomic mass 12. But the neutron count for a given element is not fixed: different atoms of the same element may have different numbers of neutrons. These differences account for isotopes, which have identical atomic number but different atomic mass. Carbon atoms can, for example, also have five, seven or eight neutrons, and so masses of 11, 13 and 14. Carbon-13 is stable and accounts for a little over 1 per cent of naturally occurring carbon, but carbon-11 and carbon-14 are radioactive and decay by transmutation into other elements. They are therefore relatively short-lived, being produced in nuclear reactions. In particular, carbon-14 is formed from nitrogen in the atmosphere in a nuclear process induced by collisions of cosmic rays. It decays back into nitrogen by emitting a beta particle, with a half-life of around 5,730 years, and this decay supplies the basis of radiocarbon dating.

Neutrons are the glue that binds protons together in the nucleus. In effect, neutrons and protons attract one another via a so-called 'strong nuclear force' that overwhelms the electrostatic repulsion one proton feels for another. Without neutrons, the nucleus would burst apart. It gradually became clear that neutrons are the key to beta decay too. This decay process looked peculiar: electrons are ejected from nuclei that apparently don't contain any. But Rutherford's and Chadwick's composite notion of the neutron has some validity: in beta decay, a neutron can decay into a proton and an electron, under the auspices of a second nuclear force (the 'weak force'). The electron leaves as a beta particle; the proton remains, transmuting the atom into the element one position to the *right* in the periodic table.*

It was Rutherford's intuition about the value of the neutron for experimental nuclear physics that most excited many of his contemporaries. Without any charge, neutrons could burrow into a nucleus at energies far below those required by protons and alpha particles. No accelerator was needed. Chadwick's colleague Norman Feather at the Cavendish soon showed that this was so, using neutrons to transmute nitrogen into boron. At the Kaiser Wilhelm Institute for Chemistry in Berlin (which Chadwick visited in June 1932), Lise Meitner and her assistant Kurt Philipp followed suit by transforming oxygen into carbon with neutrons.

The neutron changed everything, and at the Solvay conference on the 'Structure and Properties of the Atomic Nucleus' in Brussels in October 1933 their nature was debated intensively. Hans Bethe, another Sommerfeld protégé who worked at the University of Munich before emigrating to England in 1933, has asserted that everything in nuclear physics before Chadwick's discovery in 1932 was 'prehistory'.† The real history, according to Bethe, started with the neutron.

* A third particle is emitted in beta decay too: a neutrino (or its antiparticle, the anti-neutrino), postulated by Wolfgang Pauli in 1930, who wanted to call *it* the neutron. Enrico Fermi revived the idea in 1934 in his theory of beta decay, when he gave this neutral, ultralight particle its present name, 'little neutron'. The neutrino was not definitively detected experimentally until 1956.
† Bethe went to Cornell University in 1935 and worked on the Manhattan Project. He was awarded the Nobel Prize in Physics in 1967 for his work on nuclear fusion and the formation of elements in stars.

The world set free?

It was also in 1932 that the itinerant Jewish Hungarian physicist Leo Szilard, sometime collaborator of Einstein in Berlin, first read H. G. Wells' book *The World Set Free* (1914). Here Wells looked into the future forecast by Soddy, Rutherford and Aston, in which humankind had learnt how to liberate nuclear energy. Wells wrote of a war between England, France and America on one side, and Germany and Austria on the other, beginning in 1956. It would use what Wells called 'atomic bombs', which would destroy all the major cities of the world.

Szilard had gone to Berlin in 1919, where he studied under Laue, Planck and Einstein before becoming Laue's assistant and then a lecturer at the university. But in April 1933 he fled to Vienna, leaving just before the train became crammed with refugees from the Nazis. By September Szilard was in London, just another unemployed refugee himself. Here he read in *The Times* about a talk Rutherford had delivered to the annual meeting of the British Association for the Advancement of Science on 'breaking down the atom' and the 'transformation of the elements'. Rutherford had mentioned the recent experiments by his colleagues John Cockcroft and Ernest Walton at the Cavendish, who had used a particle accelerator to fire protons at lithium atoms and split them into fragments, releasing a tremendous amount of energy. Rutherford doubted that this was a practical way to generate energy, rather hastily asserting that 'anyone who looked for a source of power in the transformation of atoms was talking moonshine'.

Szilard had a healthy disdain for proclamations that such and such was impossible, all the more so if they came from 'experts'. Could Rutherford be proved wrong?

It was while walking through the London streets that Szilard saw how. The answer would surely have occurred to any nuclear physicist, Rutherford included, sooner or later, and the neutron was the key. Bothe's experiments had demonstrated that nuclear reactions could produce neutrons. Feather, Meitner and Philipp had shown that neutrons could induce them. What if there was an atomic disintegration that could both be triggered by the absorption of a neutron, and expel a neutron during the decay? Then the process could, once triggered, be sustained spontaneously, all the time releasing energy. If it happened slowly, this would create a continual source of heat which might be tapped for the generation of power. But such a self-perpetuating nuclear reaction could become a runaway process if the number of neutrons produced exceeded the number stimulating their production.* What if there was an element that absorbed one neutron and decayed to emit two? Then it would develop a chain reaction, a cascade of disintegrations that would bloom in an instant into a tremendous output of energy: an explosion. In his mind's eye, Szilard saw the apocalyptic fury of Wells' atomic bombs. As writer Richard Rhodes has put it, 'time cracked open before him and he saw a way to the future, death into the world and all our woes, the shape of things to come'.

On 4 July 1934 – coincidentally the day Marie Curie died – Szilard filed a proposal with the British Patent Office for a method to harness nuclear energy based on a chain reaction of neutron-induced atomic disintegration. He never imagined that the vision would be his alone, and it wasn't.

However, when that same year Max Planck dangled before Goebbels the promise of 'revolutionary innovations' in nuclear physics, he had as yet nothing so concrete in mind. There was no talk of weapons, nor even a clear intimation of a source of unlimited power. All the physicists understood that atomic energy might be tapped one day, but most concurred with Rutherford that the prospect was indefinitely remote. Planck's gambit was not much more than a ploy to garner state support for basic science: to wring money from the Nazis.

The German Marie Curie

Like Marie Curie, Lise Meitner always knew that she had to achieve more than her male colleagues if she was going to forge a career in science. When she arrived from Vienna to study in Berlin in 1907, Prussia still did not admit women to its universities. That changed the following year, but attitudes did not. Many academics were convinced that women would undermine the social and intellectual character of the universities. Planck, who Meitner revered for his

---

* It's not quite so simple, as Szilard realized: the liberated neutrons would have to be captured by other nuclei with adequate efficiency, so that they don't just escape from the radioactive material. The efficiency of neutron capture determines the amount of substance needed to sustain the chain reaction: the critical mass.

'inner rectitude', struggled to accept the idea: while a woman might be admitted into his discipline if she 'possess[es] a special gift for the tasks of theoretical physics and also the drive to develop her talent', he felt that 'this does not happen often'. In general, he said, 'Amazons are abnormal, even in intellectual fields . . . Nature itself had designated for woman her vocation as mother and housewife.' Marie Curie was familiar with such wearisome views; when Einstein called Meitner 'our Madame Curie', he was saying more than he intended.

Otto Hahn, based at the KWIC after his return from Rutherford's laboratory in Montreal, met Meitner in September 1907 and they decided that, sharing an interest in nuclear chemistry, they would work together. But women were not permitted inside the institute, allegedly because its director Emil Fischer was convinced they would set fire to their hair in the laboratories. As a compromise, Meitner was given a room in the basement, but forbidden, rather symbolically, to come upstairs even to talk to Hahn. The partnership was immediately productive, and by the end of 1908 the pair had published several major papers in the field. Even Rutherford had heard of Meitner when he visited Berlin on his return journey from the Nobel ceremony in Stockholm. Clearly, however, he did not know much else about her. 'Oh, I thought you were a man!' he confessed with his characteristic Antipodean bluntness.

By the 1930s Meitner had established her credentials sufficiently to join Marie Curie and her daughter Irène as an honorary man in the ranks of physics, her sex tacitly ignored. There the three of them sit, surrounded by starched collars and ties, in a photograph of the Solvay conference in October 1933, Meitner in particular looking small and frail, her eyes directed to another part of the room. But by then she will have had other things on her mind besides nuclear physics.

In 1934 Meitner and Hahn began to study neutron bombardment of uranium at the KWIC, where Hahn was now the director. This was basic science, not a quest to make nuclear power practical. They wanted to understand the sequence of transmutations that uranium underwent, stimulated by Fermi's claim to have (perhaps) found new elements heavier than uranium by neutron irradiation. If uranium were to absorb a neutron and then undergo beta decay, it would gain a proton and so become the next element in the row of the periodic table: element 93, which was not known to exist in nature. In that

Lise Meitner, seated second from right at the 1933 Solvay conference. Irène Joliot-Curie is seated second from left, and her mother Marie Curie fifth.

year, Fermi suspected he might have found it, and perhaps even element 94 too. These so-called trans-uranic elements were given provisional names based on their presumed chemical similarities with the elements that would sit above them in the periodic table: eka-rhenium, eka-osmium. To sift through the products of these nuclear reactions required chemical adroitness, which was Hahn's forte. To understand them, one needed Meitner's physics.

Fermi was wrong about his trans-uranics, but the principle of making elements by neutron capture was sound. Hahn and Meitner, assisted by a young German chemist named Fritz Strassmann, began to gather evidence for new types of radioactive substances created from uranium: maybe other isotopes of that element, maybe new elements in the uncharted territory beyond. There was something here, but it was hard to interpret.

Strassmann joined the KWIC in 1929 as a student intending to seek subsequent employment in industry, but he soon decided that he would rather stay and do fundamental research. His career after 1933 shows how difficult it could be for a young researcher not protected by wealth or status openly to oppose the National Socialists. He despised the regime and his refusal to join any Nazi organizations prompted his resignation from the Nazi-controlled Society of German

Chemists. As a result he was blacklisted from jobs in academia and industry, and denied promotion or proper pay at the chemistry institute. Consequently, he was pitifully poor and malnourished, and considered himself lucky when Hahn and Meitner managed to find an assistantship for half-pay in 1935: 'I value my personal freedom so highly that to preserve it I would break stones for a living', he attested. His resistance to the Nazis never flagged; during the war, he and his wife hid a Jewish friend in their apartment. In retrospect Strassmann looks more heroic than most of his illustrious colleagues.

Splitting apart

After Lise Meitner's flight from Germany in July 1938, Hahn and Strassmann continued the uranium studies. But without Meitner's expertise they had difficulty interpreting what they saw. They found that uranium could be transformed by neutron bombardment into three radioactive substances that seemed chemically similar to barium, and which they therefore concluded must be isotopes of radium (which shares a column with barium in the periodic table). That, however, implied two alpha particles must be emitted at once from uranium, which had never before been seen. They wrote to Meitner, now in Stockholm, who replied that it did not seem credible. In fact, these forms of 'radium' resisted all attempts to separate them from barium itself – as if they were indeed nothing other than barium.*

But that was even more absurd. Barium (element 56) had barely half the mass of uranium (element 92). There was a consensus that transmutations happened only a little at a time: a radioactive decay would turn one element into another very nearby in the periodic table, either by losing two protons (alpha decay) or gaining one (beta decay). You couldn't get straight to barium from uranium – could you?

Yet Hahn and Strassmann were running out of other explanations for their weird findings. On 19 December Hahn wrote to Meitner in Stockholm:

* The standard chemical technique for separating decay products was to mix them in solution with known, lighter elements that could be precipitated as insoluble salts by adding the right ingredient: barium was precipitated by adding sulphate, for example. An element chemically similar to the one precipitated would come along with it, usurping some of the same spaces in the crystal lattice of the salt.

Perhaps you can suggest some fantastic explanation. We understand that [uranium] really *can't* break up into barium . . . So try to think of some other possibility . . . If you can think of anything that might be publishable, then the three of us would be together in this work after all.

At Christmas time Meitner, on holiday in a quiet Swedish village, discussed the peculiar results with her visiting nephew Otto Frisch, another exile from Germany who was now working in Copenhagen. They came to a conclusion that now seems inevitable but which then contravened all prevailing wisdom about nuclear transmutation. The uranium nuclei, they decided, had indeed been more or less split in half. Searching for a name for this process, Frisch recalled the division of living cells, and borrowed the biological term for it: uranium underwent nuclear fission. Frisch told Niels Bohr in Copenhagen, who was about to leave for a conference in the United States, and so Bohr brought the news to scientists across the Atlantic.

And here was the crux: in splitting apart, the uranium atoms also emitted neutrons, as demonstrated in early 1939 by Frédéric Joliot-Curie in Paris. Neutrons in produced neutrons out: here were the ingredients for Szilard's chain reaction.

There was a catch. Like all elements, uranium has several isotopes differing in the number of neutrons their nuclei contain. By far the most abundant form is uranium-238, which constitutes over 99 per cent of natural uranium. But the form of uranium that underwent fission with the emission of more neutrons was the rarer of the two main isotopes, uranium-235, present in natural uranium at a level of just 0.7 per cent.

What's more, the neutrons that induce fission of uranium-235 are *slow* neutrons – they move at speeds of a few kilometres per second, about the same as the speed of most gas molecules at ordinary temperatures, which is why they are also called thermal neutrons. But the neutrons emitted by uranium fission are *fast*, with speeds of many *thousands* of kilometres a second. So they need to be slowed down to develop a sustained chain reaction.

Enrico Fermi had discovered serendipitously in 1934 how to slow down fast neutrons: one needs a so-called moderator, a substance that will absorb some of the neutrons' energy. Fermi found that he could enhance the radioactivity of uranium by bombarding it with fast neutrons if he placed paraffin wax between the neutron source and the target. The paraffin slows the neutrons so that they can be captured by uranium nuclei, causing them to decay.*

Paraffin, a substance made of carbon and hydrogen atoms, is an effective moderator because neutrons can transfer some of their energy efficiently to hydrogen atoms. Hydrogen nuclei – lone protons – have essentially the same mass as a neutron. So when a neutron hits a hydrogen atom, the atom is light enough to recoil from the impact and absorb some of the energy. In contrast, a neutron strikes a heavy atom as if it were a solid wall, bouncing back while relinquishing hardly any of its energy. Other light atoms can also act as neutron moderators; one of the most effective is carbon, which may be used as a moderator in the form of graphite. Heavy hydrogen – the isotope called deuterium, with a nucleus containing a proton and a neutron – is more effective than normal hydrogen, for which reason heavy water (enriched in deuterium) was also soon identified as a potential moderator. These two substances – graphite and heavy water – became the two favourite candidates for a neutron moderator that could sustain a chain reaction of uranium decay in a nuclear reactor.

The release of energy in this chain reaction can be speeded up if the fissile component, uranium-235, is made more concentrated. The nuclear physicists calculated that to produce a runaway cascade – a nuclear explosion that releases the energy almost instantaneously – requires more or less pure uranium-235. No one knew how to separate it from the much more abundant isotope uranium-238, however – the two isotopes are chemically indistinguishable.

---

* These were the experiments described earlier in which Fermi was hoping to make elements heavier than uranium by neutron capture. What he did not realize is that his slowed neutrons were in fact causing *fission* of uranium – exactly what Hahn and Strassmann reported four years later. It was a possibility that almost no one envisaged. One person did, however: a German chemist named Ida Noddack, who, like Meitner and the Curies mother and daughter, made significant contributions to nuclear science in an almost wholly male environment. But Noddack did not make her argument for fission very persuasive in 1934 – indeed, when her claim was submitted to *Naturwissenschaften*, Paul Rosbaud considered it 'pretentious' and a manuscript 'exactly of [the] sort we don't like in scientific publications'. Rosbaud did not hold Noddack in much regard, and considered it fortunate that she did not receive the Nobel Prize for which she was nominated three times.

Although such details remained unclear, in April 1939 the chemist Paul Harteck and his assistant Wilhelm Groth at the University of Hamburg decided to inform the Reich Ministry of War of the awesome power that could be unleashed from the uranium nucleus. 'We permit ourselves', they wrote,

> to direct your attention to the newest development in the field of nuclear physics, for in our estimation it holds a possibility for the crea-tion of explosives whose effect would be many times greater than those presently in use . . . in case the means for creating energy in the manner sketched above becomes a reality, which is entirely within the realm of the possible, the country that first makes use of it would, in relation to other nations, possess a well-nigh irretrievable advantage.

One might imagine that, with war in Europe looking almost inevitable, the Nazis would have considered the discovery of Hahn and Strass-mann too sensitive to disclose to the international scientific commu-nity. But Paul Rosbaud encouraged the researchers to publish an account of their work in *Naturwissenschaften*, before anyone saw fit to suppress it, and helped to rush it into print in January 1939 so that foreign scientists might become aware of it. (As we've seen, Bohr knew already.) For good measure Rosbaud told the British expert on particle accelerators, John Cockcroft, about it on a visit to Cambridge, apparently with Hahn's blessing. At the end of April the Nazi govern-ment decided that nuclear research should thenceforth be kept secret. But it was too late. In August, Einstein, Szilard and Edward Teller, another Hungarian Jew in exile from Germany, wrote a letter to President Roosevelt warning of the feasibility of making an atom bomb.

Harteck, a specialist on isotope separation, later confessed that it was not so much through patriotic duty that he and Groth sent their memo to the military, but simply because they wanted funding for their research and hoped that the army would provide it. They were certainly not disclosing any well-kept secret: anyone could see the implications of fission for themselves in an article published, again in *Naturwissenschaften*, in June by Hahn's assistant Siegfried Flügge, titled 'Can the energy content of atomic nuclei be utilized in technology?' At any rate, the letter of Harteck and Groth reached the German

Army Weapons Bureau, which decided to convene a group of special-ists to decide what should be done about uranium. They became dubbed the *Uranverein* – the Uranium Club – and they met for the first time in early September, led by Kurt Diebner, a nuclear physicist who acted as the bureau's specialist adviser on explosives. The *Uran-verein* decided that research on this new potential source of energy and military supremacy should begin right away, and at a meeting of 26 September a suitable location was identified for this work. Where else should it be but the Kaiser Wilhelm Institute for Physics in Dahlem, headed by Peter Debye?

9 'As a scientist or as a man'

On 16 September 1939, two weeks after Britain declared war on Germany, Peter Debye received a letter from Ernst Telschow, general secretary of the KWG, declaring that the Kaiser Wilhelm Institute for Physics was thenceforth to be deployed 'for military technological ends and activities related to the wartime economy'. The clear implication was that it would use its formidable technical facilities to investigate nuclear science and the possibility of liberating energy from uranium.

Such sensitive research could not be left in the charge of a foreigner, even one as deeply immersed in German science as Debye. On the same day that Debye received written notification of the impending changes at the KWIP, Telschow visited him in person to deliver an ultimatum from Rudolf Mentzel of the Reich Science Council: he must either become a German citizen, giving up his Dutch passport, or he must resign.

Debye was not prepared to relinquish his citizenship. But neither would he resign. Instead, he simply refused the demand. As he explained in a letter two days later to KWG president Carl Bosch:

> In response to Dr Telschow's first question, I replied that I did not want to give up my Dutch nationality and that he could regard that answer as definitive. As for the second part of his statement, I pointed out to him that, as he knew, during the last two years I have been given the opportunity to take up a new post, and that each time I have declined the offer without claiming any compensation. Under pressure now, I do not intend to behave differently. I thus refuse to resign my position and must demand that any initiative in that direction will have to come from my superiors. I added that I will not let people say that I ran away from it.

'There was no question of surrender', he told Laue early the following year.

Very well, Telschow called subsequently to tell him: Mentzel was prepared to discuss the situation. They met in early October, when Mentzel made it clear that discussion did not mean negotiation. He advised Debye that if he would not comply then he would have to be removed from the directorship and, as Debye put it, should 'stay at home and write a book'. Mentzel added that there would be no more funds for any research at the institute except that directed towards 'knowledge and approaches for military objectives'.

Debye regarded the KWIP as an embodiment of his life's work, and he was not going to give it up easily. That April he had accepted an invitation to deliver a series of prestigious lectures at Cornell University, and now he saw this as a way to stave off any final decision on his directorship. He arranged with the REM and the Army Weapons Bureau (which was assuming control of the KWIP) to take a six-month leave of absence from his post in order to go to America. After that, his position at the institute could be reconsidered. It was agreed that space would be made for him on his return to conduct experiments at very low temperatures, while in the meantime Kurt Diebner would become administrative director 'at' (explicitly not 'of') the institute.

On 7 October Debye wrote to Tisdale of the Rockefeller Foundation in New York to explain the situation:

> Until now the institute has been dealing with purely scientific research only. I have been informed that the government itself wants from now on to decide the kind of questions to be treated in the institute and does not want that this shall be done under my directorship, because of my Dutch nationality. As I am not willing to change my nationality, I agree with the government that for the time being I cannot act as director. As a result of an interview between the leading director of the governmental department and myself, which took place the day before yesterday, we came to the following agreement. I do not resign, instead a leave of absence will be granted for the time of the occupation of the institute during which I will be free to direct my activities, as I think best. During this time my salary will be paid as usual . . . I am very sorry that for a lapse of time, of which the duration cannot be evaluated in this moment, my work in the Max Planck Institute has ended.

Debye clearly felt an obligation to keep the foundation informed of what was happening at the institute they had paid for, but he also hoped to enlist the help of his potential allies in New York. The Rockefeller indeed provided him with unflagging support: Tisdale and other staff stepped in on several occasions during the first few years of Debye's residence in America to help with visa issues and other inconveniences, such as getting clearance for Debye's scientific apparatus to be shipped to Cornell.

Debye never made any suggestion that he was leaving Germany because of moral scruples about the regime. He undoubtedly disapproved of the military takeover of the KWIP, but there was never an indication that he left because of the *nature* of the work that would be undertaken there. Indeed he probably shared the view of the German physicists that uranium research raised a host of interesting scientific challenges. Debye was invited to the first meeting of the *Uranverein*, and might well have joined the club if the Nazis had been less intransigent about the issues of nationality and autonomy. Since he was a more able experimentalist than either Diebner or Heisenberg (who later led much of the research), it is quite conceivable that the German uranium work would have progressed faster if he had been put in charge.

However, Debye's departure from Germany later came to be seen as exemplifying his 'resistance' to the Nazis. It was presented as a flight from the fascist regime, in fact almost as a mission to warn the Allies of the developing nuclear threat in Berlin. A 1951 report in the *New York Times* put it like this:

Dr Peter J. W. Debye, a Dutch chemist and winner of the Nobel Prize in 1936, had been working at the Kaiser Wilhelm Institute at Berlin. Abruptly he was informed that his laboratory was needed 'for other purposes'. He made a few discreet inquiries and learned that a large part of the institute was turned over to uranium research. He fled Germany and came to the United States . . . Upon his arrival he notified his fellow scientists about the new emphasis the Germans were placing on nuclear research. 'His tidings', Dr Pegram [George Pegram, who headed early work on the Manhattan Project] said, 'started a race between our scientists and the German. From then on we worked day and night in a race to get ahead of the Germans.'

It is a very short step from here to the conclusion that Debye came to America to inform on the Nazis and thwart Hitler, rather than because he was seeking a way to postpone a final severance from his Berlin institute. That became the 'Debye story', which went essentially unchallenged until his death. Yet despite the claims of his accusers today, Debye didn't exactly fabricate any part of it. He seems to have been content to let the myth evolve on its own.

Sailing to America

The German authorities were well aware what Debye's departure might mean. Telschow made his concerns clear in a letter to Adolf Baeumker, chancellor of the Reich Academy of Aviation Research:

> In my view there is also the possibility that Prof. Debye . . . may now decide to abandon Germany . . . The Kaiser Wilhelm Society will make efforts to find other possibilities of employment here in Germany after Prof. Debye has concluded his lectures in Ithaca, and it would be welcome if the Academy for Aviation Research could be deployed in a similar way.*

In December Debye and his wife Mathilde went to Maastricht, ostensibly to visit his sick mother but in fact to discuss plans for how to leave the country. With this in mind, Debye borrowed money in dollar currency from his mother. He was supposed to meet Telschow later that month to sign the contract of leave, but Telschow found when he arrived at the institute that Debye was once again ahead of the game, and had left already.

It was not exactly an escape, given that he had negotiated official permission, but the Debyes took care to ensure that the departure should not seem to have an air of finality. On 15 January 1940 Debye travelled from Munich across the Brenner Pass to Milan and then Genoa, where he picked up more money transferred by his mother

---

* In seeking for solutions, Baeumker wrote to Ludwig Prandtl at Göttingen, mentioning that he understood Debye to have been granted leave of absence 'for the duration of the war'. The implication that the war might be over in six months or so reflects a common view in Germany at that time, and it is worth bearing in mind that Debye might have considered that to be possible too.

and boarded the *Conte di Savoia*, setting sail for New York on the 23rd.

Debye's son Peter, then working for a doctorate at the University of Berlin, had already gone to America in the summer of 1939 on a holiday with friends in Ohio, and the outbreak of war prevented his return. (It has been suggested by the Debye family that this was never intended anyway – that Peter's departure was the initial move in a planned exodus.) Debye intended that Mathilde would join him at Cornell too. This required some ingenuity. Although she could not obtain an American visa directly (she had, on marrying Debye, become a Dutch national), she could travel to Switzerland where this might be arranged. It turned out to be far from easy: at first she was denied, and only after the intervention of the presidents of Cornell and the Massachusetts Institute of Technology did the US State Department finally grant Mathilde permission to travel. She left Lisbon in December 1940, sailing to Cuba just before Christmas and arriving in the United States in January 1941 to join her husband in Ithaca.

Debye's departure was painful; he must have suspected that it would be hard, if not impossible, to return. 'I had to give up all these beautiful laboratories which I had built', he recalled wistfully in 1964. 'It cost a few million, you see. I had everything the way I wanted it – these high voltages and so on and at Cornell they had nothing.'

A society at war

Debye's abrupt departure left the German Physical Society without a chairman. His deputy Jonathan Zenneck stepped in until a new leader could be selected – a process that was inevitably politicized. The Nazi sympathizers on the committee, Stuart, Schütz and Orthmann, favoured the experimental physicist Abraham Esau at Jena, who was head of the physics section of the Reich Research Council and had been a party member since 1933. But the moderates were able to engineer the election of Carl Ramsauer, an industrial physicist who, like Bosch at the KWG, could be considered politically sound while being somewhat insulated from state interference. Ramsauer was a former student of Lenard, but shared none of his mentor's rabid anti-Semitism. He was a conservative and nationalist but not a party member. Ramsauer set the DPG on a course of partial, voluntary alignment. He implemented the Führer principle, as stipulated in the society's new statutes of 1940 over which Debye had dragged his heels, and he acknowledged that the DPG had a duty to contribute to national defence. But he exercised that duty by strengthening ties with what had now become the German military-industrial complex, which was powerful enough to operate on its own terms and set its own (self-interested) priorities rather than being dictated to by the government.

This strategy gave Ramsauer a strong hand for obtaining more funds for physics. He emphasized the discipline's importance for national security at every opportunity, and in 1942 the DPG felt bold enough to complain to Rust that the fiction of a 'Jewish physics' had been so damaging that 'German physics has lost its former supremacy to American physics and is in danger of continuing to lag behind'. True, the scientists didn't expect a favourable response, and in the event they got none, owing to Rust's notorious sloth and inefficiency. So the DPG began to exploit the greater interest in their work shown by other factions in the Nazi bureaucracy, particularly the head of armaments Albert Speer and Air Marshal Hermann Goering.

Nuclear research was particularly valuable for gaining traction with the authorities. In March 1942 Ramsauer told General Georg Thomas, head of the Military Economic and Armaments Office, that 'the large-scale research [in America] in the area of nuclear disintegration could one day become a great danger for us'. As a result of these petitions, says historian Klaus Hentschel, there was an 'increasing acceptance of physics by the Nazi authorities' during the later years of the war. Goebbels himself was ready to admit the wounds that Nazi policies had inflicted on their scientific capacity (while finding a suitable scapegoat), writing that

> our technical development both in the realm of submarines and of air war is far inferior to that of the English and the Americans. We are now getting the reward for our poor leadership on the scientific front, which did not show the necessary initiative to stimulate the willingness of scientists to cooperate. You just can't let an absolute nitwit [meaning Rust] head German science for years and not expect to be punished for such folly.

As a result of this belated enlightenment, science became rather well supported as the war progressed, even while the German economy as a whole became increasingly parlous. The KWG saw its funds swell from a budget of 5.5 million Reichsmarks in 1932 to 14.3 million in 1944. Ramsauer also won exemptions from active military service for many physicists, arguing that while the army could surely bear 3,000 fewer men, '3,000 more physicists could perhaps decide the war'. By the time he made this appeal, most people knew that the war was already decided – but naturally that only enhanced the impact of such promises. The petition was rather too late, however: of 6,000 scientists who the Nazis tried to recall from combat in 1944, only 4,000 returned, the others being either already dead or untraceable. Besides, some of these scientists decided not to waste their efforts on a war that was already as good as over, and so they used military contracts to pursue what were really just academic studies, while claiming that these were of profound military significance. On this basis, Max von Laue justified publishing a book on the theory of diffraction which could not be of the slightest value to the military. 'If someone wanted to research persistently through the files of the final years of the war', Laue wrote in 1946, 'he would notice that absolutely everything conducted in science was "decisive for the war effort".'

Debye in America

Debye's arrival at Cornell was regarded with much suspicion. Why had he left it so late to leave Germany, unless he felt some loyalty to the country? At the end of August 1940, Samuel Goudsmit wrote to the FBI to say that

> Some of my colleagues think that [Debye's] new position here may bring him into contact with scientific defense work and that he may have an influence upon the choice of personnel for that work. They fear that he is not reliable. My own opinion is that these suspicions are primarily caused by professional jealousy. I hope that I am right. Nevertheless the case seems important to me. Debye is such an outstanding man in his field with broad practical experience that it would be a serious handicap if, in an emergency our country would

be unable to use his valuable knowledge because of unfounded suspicions. It seems in any case highly advisable to make sure just where he stands.

It is possible in principle that Debye could indeed have intended to spy for the Nazis in America, and there have been recent speculations to that effect. But this would contradict just about everything else we know about his attitude to Hitler's regime, and indeed about his character and conduct both before and after the war. The idea lacks not only evidence but logic and psychological plausibility.

The FBI could not have known that, and it launched an extensive investigation on Debye, interviewing many of his peers and colleagues. The responses were strikingly polarized. People who had been close to Debye in Germany tended to speak well of him, to recommend him as a man of integrity who did not dabble in politics but lived only for science. But a few of those whose relationship was more distant were wary and critical. Several of them were Jewish émigrés; Goudsmit himself told the investigators that although he had not seen Debye since 1931, he was 'suspicious of him, but this has no basis in facts'. The Polish radiochemist Kasimir Fajans, who left Germany in 1935, stated that Debye would turn a blind eye to anything problematic when expedient; another Jewish Pole, the physicist Roman Smoluchowski, who escaped from Warsaw in 1939 to come to Princeton, pronounced him 'extremely mercenary as a scientist'. The German physicist Rudolf Ladenburg, who had worked at the KWIPC and the University of Berlin before emigrating to Princeton in 1932, concurred, saying that Debye was 'not loyal even to the field of science, where money was involved'. Russian-born physicist Gregory Breit at the University of Wisconsin, who had been in the United States since the 1920s, attested that he could not rule out the possibility that Debye was working for the Nazis against his will. Others mistrusted him apparently because of his very brilliance. Wolfgang Pauli warned that Debye 'could not be trusted' and claimed that he was 'in all probability very sympathetic to the German cause' – an accusation characteristic of the abrasive Pauli in being both withering and unfair. In America Debye rarely spoke about his attitude to Hitler's regime – an unidentified FBI interviewee stated that he 'has no emotional reaction whatever to the whole Nazi question' – but there seems little reason to doubt the testimony of Harvard chemist Frederick Keyes in 1940 that 'Debye detests thoroughly all about Hitler and the Nazi government.'

Potentially most damaging was the testimony of the most authoritative of the Jewish German physicists, Einstein. Although he never made it explicit, Einstein evidently did not think highly of Debye's morality. As the FBI report attests,

> Einstein advised that he has never heard anything wrong concerning Debye but that he knows the man well enough not to trust him; that he Einstein would accept things that Debye says as a scientist as being true but would not accept things that Debye says as a man as necessarily being true. Einstein continued that Debye is a very shrewd man of extraordinary intelligence, very versatile and having extraordinary ability to reach his goals and knows what to do to obtain immediate and personal advancement. Einstein said that he believes Debye is not a person of high loyalty and will use anything for his own advantage. Einstein stated that Debye acted very suspiciously abroad and did not act as a Dutchman. In explanation of this, Einstein said that Debye's colleagues abroad had been persecuted since 1933 and that he [Debye] in no way tried to help them and did not attempt to aid them in securing positions elsewhere.

This is a peculiar statement, not least because it is incorrect in the last respect. One can argue about the moral basis for his actions, but Debye evidently *had* helped persecuted colleagues to leave Germany and find posts elsewhere, including his assistants Heinrich Sack and Willem van der Grinten, and of course Lise Meitner. That Einstein was apparently prepared so definitively to make these claims without knowing exactly what had transpired in Germany since 1933 does him little credit.

Einstein went on to sow seeds of doubt about Debye's loyalties:

> Einstein said that he does not believe Debye's work [at the KWIP] concerns military affairs but that Debye is capable of performing such work. He said that Debye may be all right but that if Debye's motives are bad he is a very dangerous man. He also stated that Debye would be a good man for [German] espionage work as he has the facility of organization to perform such work. He said that it was his unbiased

opinion that Debye should not be trusted with military secrets of the United States government, unless it has first been ascertained that Debye had severed all relations with German officials.

It isn't clear why Einstein had so jaundiced a view of Debye. When they first met around 1917, Einstein referred to him in only favourable terms, an 'unspoiled soul' whose scientific abilities he held in great regard. Gijs van Ginkel, the former managing director of the Debye Institute at the University of Utrecht, suspects that the relationship might have soured in the 1920s because the Debyes knew through a mutual friend in Zurich all the details of Einstein's rather shabby treatment of his first wife Mileva Marić, and that Mathilde Debye made no secret of her disapproval. Debye's opinion of Einstein seems to have declined too – his sister Caroline stated in 1970 that 'he found Einstein in fact to be a windbag, a person who had been made too much of'. Debye may have resented Einstein for squashing his chances of gaining a professorship at Leiden University in 1912 by nominating Paul Ehrenfest instead.

In any event, Einstein's distrust caused Debye some discomfort, especially after Einstein received a strange letter from Europe in the spring of 1940. It was brought to his home in Princeton by a British intelligence agent, having been spotted in the mail by the censors. The letter was from a man in Switzerland who Einstein didn't know, named Feadler (as the FBI transcribed it) or some such. (Possibly its author was one Hans-Werner Fiedler, on whose doctoral thesis under Heisenberg Debye had acted as an assessor.) This letter warned that Debye was close to Hermann Goering (which was true) and had quite possibly come to the US for a 'secret purpose', which the author hoped Einstein might establish.

That last suggestion was mere conjecture. Nonetheless, 'since I had no possibility to investigate the statements made in this letter', Einstein later explained,

> I gave the information to one of my colleagues here with whom I am befriended. This was self-evident duty . . . Under the circumstances, I could not take the responsibility to throw the letter into the waste-basket.

He passed the letter on to the palaeographer Elias Avery Lowe at the Princeton Institute for Advanced Study, who in turn forwarded it to a 'Jewish academic' at Cornell. The authorities there advised Debye of the charges.

Unsurprisingly, Debye wrote rather stiffly to Einstein. 'Those suspicions are entirely groundless', he said. 'I have left Germany because I was asked to change my Dutch citizenship into a German citizenship. [I] decided some months ago that under no circumstances would I return to Germany.' According to the president of Cornell, Edmund Day, Debye dismissed the 'Feadler' letter as 'symptomatic of the kind of hysteria which we are doubtless in for'. Day assured Warren Weaver of the Rockefeller Foundation that he 'has every confidence that Debye is honest and loyal'.

Einstein mentioned the 'Feadler' letter in his interview with the FBI, but Robert Ogden, dean of liberal arts at Cornell, considered it 'the result of Jewish prejudice' – that is, of the bitterness that those forced out of Germany felt against others who, unaffected by the Nazi edicts, seemed heedless of their plight by remaining. Whether or not this accounts for Einstein's feelings, he made them plain enough. When a reception dinner was held at Princeton to honour Debye in June 1940, Einstein 'declined to attend', according to Weaver. At that event, Weaver was told by the mathematician Oswald Veblen that 'some of Debye's colleagues obviously considered that he had been too tardy in coming to his conclusion [that he] could no longer associate himself in any way' with Germany.

Not all Jewish émigrés judged Debye harshly. James Franck, who would probably have headed the KWIP instead of Debye had he not been expelled from Göttingen, asserted to the FBI that 'Debye is a man of high character and high ideals, he is totally trustworthy and would be totally loyal to the American government.' But it is notable how differently Debye appeared when seen from afar, compared with the impression he tended to make at close quarters. Something about his manner in the world left the likes of Einstein and Goudsmit uneasy in a way that can't be ascribed simply to professional jealousy.

It took the American authorities the best part of four years to decide that Debye could be trusted. In April of 1944 the Army Service Forces stated that it saw no reason why he should not be permitted to participate in classified military research. By that stage it hardly mattered.

Life in wartime

Debye had already been contributing to the Allied war effort as best he could. As the FBI report noted, he wasted no time in telling scientists in America what the German physicists were up to in Berlin. Two weeks after his ship docked in New York, he met with Weaver to explain the situation. As Weaver put it:

> The army has made this move [at the KWIP] because of their hope (which Debye considers quite misplaced) that a group of German physicists working feverishly with Debye's excellent high tension equipment will be able to devise some method of tapping atomic or subatomic energies in a practical way; or will hit upon some atomic disintegration process which will furnish Germany with a completely irresistible offensive weapon. That this is indeed the army's hope and plan is supposed to be a great secret, and Debye himself is not supposed to know this. Nor is anyone supposed to know the German physicists who are entering into this scheme, although Debye has already told us who these are. Debye says that these German physicists very definitely have their tongues in their cheeks. With Debye they consider it altogether improbable that they will be able to accomplish any of the purposes the army has in mind; but, in the meantime, they will have a splendid opportunity to carry on some fundamental research in nuclear physics. On the whole Debye is inclined to consider the situation a good joke on the German Army. He says that those in authority are so completely stupid that they will never be able to find out whether the German physicists are or are not doing what they are supposed to do.

Notice that Debye was not trying here to undermine Germany's military research programme. Rather, he passed on this information with the view that it was of no real consequence, believing that the nuclear work was a waste of time and indeed little more than a joke. He told an American magazine for Dutch immigrants at the end of 1942 that 'the chances are infinitely against the development of any really new and astounding weapons, on either side in the present war. Any "new" weapons will only be developments of or improvements upon the present instruments of combat.' That is not quite what the German physicists themselves believed, as we shall see. In any event, the information that Debye freely revealed about the German uranium research provided an important stimulus for the intensive effort at Los Alamos. It was this information that Einstein (evidently trusting Debye's words and intentions at least to this extent) and Leo Szilard drew upon in their second letter to President Roosevelt in April 1940 imploring him to support large-scale research on the liberation of nuclear energy.

Debye's defenders today are understandably dismayed that his accusers can and will damn him for his actions in America whatever he did. If he had shown no interest in war research, that would have exposed lingering German sympathies. But since Debye *did* conduct defence work, he was clearly an opportunist, ingratiating himself with whoever held the reins of power.

Yet if Debye was willing to conduct war work for the Allies but not for the Nazis, doesn't that in fact show where his sympathies lay?

It doesn't show this at all. As Debye himself admitted to Einstein, he left Germany not because he objected to the military orientation foisted on the KWIP but because he could not retain his post unless he became a German citizen. Whether, had he been allowed to keep his Dutch passport, Debye would have stayed and pursued uranium research can only be a matter of speculation. While we simply do not know how ready he would have been to work for the German Army, it seems unlikely that he would have had any qualms about working on applications of nuclear energy per se. When asked by a Dutch newspaper in 1948 whether he had been involved in the American nuclear research, he evasively replied: 'I could have done so, of course – we had already carried out researches in this area in Berlin – but as a Dutch citizen I did not consider it right. Besides, I had found other work at Cornell.' This is a highly disingenuous remark. Not only is it unclear why being a Dutch citizen would in itself make nuclear work 'not right', but Debye knew very well that the real reason he was not asked to join the Manhattan Project was that he had no security clearance. It is no wonder that some people, then and now, accuse Debye of arranging facts to suit himself.

Instead of nuclear physics, Debye undertook rather more prosaic assignments for the war effort. He studied insulating materials used for radar systems, partly in collaboration with Bell Laboratories, which applied for permission to involve him in 1941. The company was told in December that Debye 'should not be entrusted with any confidential navy matter'.

Most importantly from both the scientific and military perspectives, Debye helped to develop synthetic rubber, sorely needed by the military after the American sources of rubber in the Far East were cut off by the Japanese offensive. Debye figured out how, by looking at the patterns in light passed through cloudy solutions of polymers, one could deduce the average sizes of the long-chain molecules. This solved an important problem in the young field of polymer science, and Debye's results are still valuable today. He also worked on the theory of the flexibility and elasticity of these materials.

Debye's participation even in these defence projects was initially treated with almost farcical caution. He was allowed to visit defence-related establishments or labs for meetings only if accompanied by a military policeman – as if, as one colleague put it, he might otherwise storm the place and blow it up. Debye's colleagues found these restrictions both comical and deplorable. William Baker, a former director of Bell Laboratories, who worked on the radar project alongside Debye, commented that Debye 'took it with immense good humour and, of course, endeared himself so quickly to the guards that they probably could have been enlisted on his side in any venture that he wished'.

Letters to Berlin

Had Debye really renounced Germany for good? According to his grandson Norwig Debye-Saxinger,

> Accepting the Baker [Cornell] lectureship was playing for time. He felt
> he had to get an employment commitment sufficient to bring over his
> wife and daughter . . . In the US, no one who dealt directly with Debye
> was confused about his intentions: he was here to stay and determined
> to keep his son here and bring his wife and daughter over.

This claim is somewhat supported by Weaver's comment in mid-April 1940 that 'Debye has now practically definitely decided to remain in the United States.' The Rockefeller Foundation awarded him a grant of $17,000 for research at Cornell over a period of three years. On 17 June Debye signed a contract for a permanent position at Cornell, becoming head of the department. This contract made Mathilde, then still in Switzerland, eligible for a US visa. Even if Debye had not yet decided to emigrate to the United States when he boarded the *Conte di Savoia,* then, it appears that this decision was made soon enough. There seems to be little to argue about here.

But there is. The view of Sybe Rispens, reiterated in Martijn Eick-hoff's report for the Netherlands Institute for War Documentation, is that Debye maintained secret contact with the German authorities while in America, hoping to 'keep the back door open' so that he might slip back into his post the moment this became feasible. In other words, he was acting as a shameless opportunist, his apparent allegiance to the United States being no more than a matter of expediency.

It's true that Debye never formally served notice of termination for his directorship of the KWIP, and seemed keen to keep the matter open. In the spring of 1940, as the agreed six-month period of absence approached, Bernhard Rust wrote to Debye at Cornell to ask what his plans were. He could conceivably have replied that he, his wife and his son, were now going to live in America, and that would have been the end of the matter. Instead it looks as though he was intent on sowing confusion. The first communication that Ernst Telschow of the KWG received from Debye in America was a telegram of 25 July, saying 'My letters remain unanswered in the circumstances have decided to accept broader offer Cornell new letter in the post.' Debye explained in a subsequent letter that he had written previously to ask about his status at the KWIP. There's no telling if that is true.* In any

---

* In early 1946, when Debye's application for US citizenship was being considered, he told the FBI that he had never answered Telschow's letter of 1940 asking about his plans, and that he'd heard no more subsequently from the KWG. This is clearly untrue. He told Telschow in 1940 that he'd not *received* any previous letters, while subsequently there was a fair amount of correspondence with Berlin, as we'll see. It is understandable that, during deliberations about his naturalization, he would wish to suppress these facts. His deception here certainly doesn't warrant the

case, at the end of August Telschow and Mentzel replied to say that they could extend his leave until 31 March 1941, but no further.

Telschow now doubted Debye's intentions. He wrote again to confirm that there was no question of resuming his Berlin directorship until after the war, and to probe more deeply: 'It is curious that about four weeks ago a report appeared in the Dutch press that you had decided to remain in America permanently. I hear the same here repeatedly in scientific circles.' It wasn't surprising that these rumours had travelled, for Debye's contract with Cornell in June had been reported in the *New York Times*. Debye's response must have been infuriating: a postcard saying that he considered it important to be clear and honest – and nothing else.

As the revised deadline approached, Telschow asked Debye again what he was intending to do. He received no reply, and on 1 April 1941 Debye's leave was cancelled and his salary suspended. Mentzel issued a statement saying that he had entered employment in the United States 'against the wishes of the government of the German Reich'. Nonetheless, Debye replied on 2 May to say, disingenuously, that he was ready to resume directorship of the KWIP 'as soon as you are again able to guarantee that I will have the possibility to fulfil the corresponding obligations according to the conditions of my old contract' – that is, as a Dutch national. He followed this with a telegram to the Foreign Office in Germany on 23 June:

> Professor Debye states that he is prepared at any time to resume the directorship of the institute under the previous conditions as soon as this is possible there. Until then he requests further leave to give guest lectures at Cornell University.

The telegram added that a more detailed letter would follow.

There can be little doubt that Debye was now playing a game with Telschow and the authorities. But to what end? One key consideration is that, while Mathilde and their son Peter were now safely in America, his daughter Mathilde ('Maida') was not. Both she and Debye's sister-

melodramatic conclusion that Martijn Eickhoff draws: that Debye 'could pull the wool over people's eyes' and 'was caught between several incompatible stories and could barely talk about his own past in Nazi Germany any more without doing violence to historical truth'.

in-law Elizabeth Alberer ('Aunt Lisi', who had largely brought up Maida) had remained living in the KWIP director's house in Berlin. Debye's grandson Norwig suggests that this was the motive for giving Berlin the impression he intended to return. As Mathilde later explained,

> Peter Debye continued the negotiations with the KWIP in order to keep drawing a salary during his absence. That has two aspects: on the one hand this was a source of income for his family members left behind in Berlin, to keep a roof over their heads in his KWIP house in Berlin; on the other hand, he gave the Nazis the impression that he wanted to return, so that they would not take action against his family.

There seems good reason to believe that Debye wanted his daughter out of Germany. But why had she and Aunt Lisi stayed there anyway? Laue commented that he found it 'strange' that Maida remained in Berlin even after her mother left for America, although in fairness it was by that time an extremely delicate matter to cross any national borders. Debye had written ambiguously to Arnold Sommerfeld on 30 December 1939 that 'Hilde [his wife Mathilde] and Maida prefer to wait here to see how matters develop' – but that may be merely cautious wording mindful of the postal censor, given that his wife took steps to leave soon after. It does seem that some plans were made for Maida to leave too: in June 1940 Mathilde wrote to Debye from Switzerland to say that their daughter was expecting to receive a visa so that she might join her in Lausanne. This authorization was apparently never forthcoming.

Yet one has to accept that this aspect of the Debye family's plans is murky, not least because in March 1942 Maida married the Moravian German Gerhard Saxinger, formerly in the Czechoslovakian army and now a German military photographer, who had previously rented rooms at the KWIP house. She was already pregnant when they married; her first child Norwig was born in August, and her second, Nordulf, a year later. Norwig suspects that this union may have sealed her decision to remain in Germany after all. It isn't clear (nor terribly relevant to Debye's own position) how she felt about that. Some have suggested that Debye's daughter was sympathetic to the Nazis, others that she had mental-health problems, although both suggestions are emphatically denied by Norwig Debye-Saxinger.

A memo for the Rockefeller Foundation in the 1940s inverts the later suggestions of Debye's grandson and wife, saying that he was not so much preserving the director's house for his daughter and sister-in-law as they were for him. Their presence, the report says, gave him a remaining toehold at the institute 'so he won't lose everything'. Whether it was a ruse or not, Debye did use his daughter's residence explicitly to verify his own intention to return. On 12 June 1941 the German Consulate in New York interviewed Debye about his plans; he told them

> that his wife was born in Germany (Munich), that his daughter still lived in Germany and, as far as he knew, worked at the Ministry of Propaganda, and that he and his son . . . had spent most of their lives in Germany . . . and were therefore very eager to be able to live and work there again.

Debye had apparently stated that he would be ready to forego his part of the director's salary but requested that the rent and maintenance of the house still be covered for his daughter and sister-in-law.*

Debye surely did have genuine concerns about Maida and her Aunt Lisi, particularly regarding their financial security (although this situation must have improved after Maida was married and her husband helped her find a job in a government ministry). And apprehensions about their safety in Hitler's state were not misplaced. When the Nazis discovered in 1941–2 that Debye had somehow managed to withdraw money from Germany, they suspected his sister and brother-in-law Hubert Niël in Maastricht. Niël was taken for questioning and imprisoned for six months. It's not obvious that Aunt Lisi and Maida faced comparable risks, but in wartime Germany no one could be too sure what tomorrow would bring.

Debye must have known that the director's accommodation would not be held in abeyance indefinitely, however much he prevaricated. It

---

* Sybe Rispens claims that the two women lived well in Berlin on Debye's 'royal salary'. Quite aside from the unfairness of that imputation, it is not even clear how easy it was to access Debye's salary at all: Mathilde struggled in late 1941 to give her sister Elizabeth access to her Berlin bank account. And Debye was unable to send them money from America, since the deteriorating relations between the United States and Germany had resulted in a freeze on bank transfers in June 1941.

was surprising that his relatives were permitted to stay there for so long after he left. When Werner Heisenberg was appointed new director of the KWIP in 1942 (see page 192), in theory he and his family could claim the house. But Maida and Lisi were still there, and Heisenberg was reluctant to force this delicate issue. The two were ordered to leave in May 1943, but they appealed to the Foreign Office, which wrote to the REM that 'It would be undesirable for cultural and political reasons for a scientist of such importance [Debye] to be able to refer later to the fact that his family had not only been expelled from their home in Germany but also deprived of any maintenance.' It was finally agreed that Debye's daughter and sister-in-law would leave in August but that they would receive an allowance of 400 Reichsmarks a month, along with the peculiar right (if Heisenberg did not object) to take fruit from the trees in the garden of the director's residence. Aunt Lisi went with Maida and her young family to live with their in-laws the Saxingers in Sudetenland, taking with them the 'emergency money' of Debye's gold Nobel medal hidden in a baby's nappy. However, as the Russians advanced through Sudetenland, Maida's family were forced to flee, becoming refugees and losing touch with Debye in the United States for the rest of the war and its immediate aftermath. Around late 1945, Debye was able to establish only that his daughter had been 'somewhere in Czechoslovakia' towards the end of the war, and they were not reunited until 1948. Maida Debye-Saxinger and her sons eventually emigrated to America; she divorced Saxinger in the mid-1950s.

Historian Dieter Hoffmann says that in maintaining contact with Germany, Debye was doing what many others did: 'it was common for scientists leaving Germany – for whatever reason – to try not to burn their bridges with their former home. There are many possible reasons for this, including family and future pension or compensation claims.' But why Debye, head of department and comfortably supported at Cornell, where he could stay for as long as he desired – and not even a German – would wish to do this is not at all obvious. Was he really hoping that he might some day go back to the KWIP to resume his unfinished research there? Would he have done so if Hitler had been victorious (which looked quite possible in 1941, until the United States and Soviet Union entered the war)? Would the lure of the well-equipped institute that he'd built up from scratch have been irresistible?

We simply don't know. Both the accusation that he was keeping the back door open and the supposition that he was stringing the Germans along to protect his family are guesses, and carry little weight either for attacking or for defending Debye. More to the point, both positions assume that he had a well-formulated plan – they succumb to the seductive fiction that humans understand and determine every aspect of their actions. It seems far more likely that, unable or unwilling to grasp the true political situation, Debye used prevarication as an ad hoc means of keeping the problem at arm's length, planning only for the coming weeks. There was never any devious grand scheme, just a refusal to accept how things stood. One can judge Debye for that as one sees fit; but there is scant reason to regard him as a straightforward opportunist.

Running away?

In the simple-minded dichotomy that makes exiles from Nazi Germany blameless and those who stayed culpable, Debye also occupies an ambiguous position. Yes, he left – but only after the war had begun, and only because he had been removed from his post. He never once mentioned moral reasons for his departure, and it's not impossible that, at least initially, he hoped to return. What do we make of that?

The truth is that we should again make very little of it. It shows neither that Debye was 'principled', beyond the strongly held principle of retaining his Dutch citizenship, nor that he was opportunistic. Like countless others as war broke out, Debye was improvising. He had not expected the ultimatum about his directorship. His primary goal was to work unmolested; since that was impossible in Germany, he reluctantly took the chance that America offered.

Besides, the moral dichotomy of stay-or-flee was more complicated. Max Delbrück had a dim view of cut-and-dried judgements on the matter:

Many nasty things have been said about those who could have left and didn't leave, like Heisenberg . . . I don't agree at all with these derogatory comments. I don't think that it was anything to my credit that I left at all. I think it was a question which could be answered one way or the other, and there is great merit on both sides . . . what is the moral argument [for] running away? It's just running away, that you take the advantage that you can run away. If you imagine that the [regime] may last only a short while, then it's important to see that some of the good people are staying.

In some ways, Delbrück said, it could be harder for non-Jews to leave Germany:

Going away without any kind of security – that means having a job somewhere else – was limited to those who had professions that were salable in another country and who had already professions or had some other ways of having private funds, or large funds that they could transfer, and could start a new life in a different country. But that was an infinitesimal part of the population . . . If you were non-Jewish and left you were certainly very suspect and couldn't expect much help from the Jewish organizations . . . Why would the fellow leave if he didn't have to? That was more the attitude really at the time. I mean I wasn't applauded for leaving, but I was suspected of leaving by having some sinister motive imputed. And rightly so. There were certainly quite a few Nazi agents who did leave posing as adversaries.

We saw earlier how, for Heisenberg and Planck, resignation was an abdication of one's responsibility as a German and a scientist. Debye did not share, or at least did not articulate, their sense of a duty to 'preserve German science'. Does that absolve him of blinkered, egotistical nationalism, or does it deprive him of an 'honourable' principle of loyalty, however misplaced?

For Debye, such questions seem to have been irrelevant. By all appearances, he made his choices for immediate, pragmatic and personal reasons. A factor in his reluctance to accept the offers of positions in the United States during the 1930s, for example, was that his wife, who was from a working-class Bavarian family and had only a perfunctory education, spoke no English and preferred to live in Germany. Her unhappiness during his time at the University of Utrecht contributed to his decision to return to Germany. However one chooses to weigh these personal obligations against any social and political responsibilities accrued from choosing to remain in Nazi Germany, one has to acknowledge that Debye's predicament was not easy. Perhaps what matters most is not which decisions he made, but to what extent he was able and willing subsequently to consider their moral dimension. There is no record that he ever spoke of such issues.

10 'Hitherto unknown destructive power'

The key historical question about the ethics of physicists working in Germany before the war commenced is how they accommodated their practices and institutions to the racist policies and dictatorial administration of the Nazi regime. But once hostilities began, the matter has tended to develop a different focus: more narrowly defined, more intimately bound up with science itself, and with implications that extend far beyond Germany. For historians examining this period, one of the crucial questions is whether these scientists were prepared and able to make a nuclear bomb for Hitler. No end is yet in sight for the controversy that this issue provokes, and at the centre of that storm is Debye's former colleague at Leipzig and his eventual replacement in Berlin, Werner Heisenberg.

The literally explosive implications of Hahn's and Strassmann's discovery of uranium fission in late 1938 were appreciated at once. At the same time that Paul Harteck and Wilhelm Groth at Hamburg told the Reich Ministry of War how the discovery might be exploited for energy and weaponry, James Franck's successor Georg Joos at Göttingen heard the experimental physicist Wilhelm Hanle deliver a paper on how a nuclear reactor – a *Uranmaschine*, uranium machine – might be devised. Joos and Hanle sent a letter explaining this proposal to Wilhelm Dames at the REM, who passed it on to Abraham Esau of the Reich Research Council. On 29 April 1939 Dames and Esau convened a meeting of specialists – the first *Uranverein* – to discuss the matter, including Joos and Hanle, Walther Bothe and Hans Geiger. Peter Debye was invited but did not attend. Exploratory research on uranium fission began at Göttingen but did not get far before the physicists there were called to military service in August.

Harteck's and Groth's missive reached the head of weapons research at the Army Ordnance Office, Erich Schumann. He was sceptical that there was anything in this wild idea, but sought the advice of his explosives expert, physicist Kurt Diebner of the Physical and Technical Institute in Berlin. Diebner talked the matter over with his assistant Erich Bagge, who had recently gained a doctorate in nuclear physics at Leipzig under Heisenberg. They brought together a second group of specialists in Berlin on 16 September to discuss the possibility of harnessing nuclear fission for military purposes. On the same day, Ernst Telschow informed Peter Debye that the KWIP was to be handed over to Army Ordnance for military research.

This second Uranium Club included Bothe and Geiger along with Harteck and Hahn. For their second meeting ten days later Bagge suggested adding his former professor Heisenberg, who soon came to dominate the group. Heisenberg first took the lead by writing a report for Army Ordnance on the feasibility of liberating energy by controlled fission in a uranium machine. Such a device, he explained, could provide a source of heat for powering tanks and submarines. Heisenberg's memo in December 1939 also pointed out that if the uranium was sufficiently enriched in uranium-235 then the chain reaction could become a runaway process, releasing all the energy at once: the fissile material would become an explosive 'more than ten times as powerful as existing explosives'.

No promises

Could enrichment of uranium be achieved? Harteck and others started to explore methods for separating its isotopes – an immensely difficult challenge, since their atomic weights differ by so little. Much of the initial uranium research, however, focused on making a reactor rather than a weapon, using heavy water as a moderator to slow the fission neutrons so that they could be captured by uranium nuclei to sustain the decay process. (Graphite was also considered as a moderator, but abandoned at the outset – see page 216.) Until close to the end of the war, Germany had access to only a single facility capable of separating heavy from ordinary water: a hydroelectric plant at Vemork in occupied Norway, which had been taken over after the invasion by the Berlin-based mining and chemicals company Auer. The first prototype reactor in Berlin, however, used paraffin wax as the moderator, as Enrico Fermi had done in his early experiments on slowing down neutrons. The work was conducted in a wooden building in the grounds of the KWI for Biology and Virus Research, next door to the physics institute in Dahlem. To deter inquisitive snoopers, it was called the Virus House.

Progress was slow. Germany was well positioned to conduct uranium research since it had access to the largest source of ore in the world, at Joachimsthal in what was then occupied Czechoslovakia. But to use the heavy metal in a uranium machine, it had to be processed: extracted and turned into plates by standard metallurgical techniques. During wartime, Germany's metal foundries had more urgent priorities.

With Debye gone and Diebner appointed by Schumann as acting head of the KWIP, the scientists there began to test reactor designs. They thought initially the best geometry would be a series of concentric shells of uranium separated by heavy water – a kind of nuclear onion. Stimulated by Fermi's work on trans-uranic elements, the physicists believed that neutron absorption by the predominant, non-fissile isotope uranium-238 would generate element 93, which should also be fissile like uranium-235. In July 1940 Weizsäcker suggested to the Weapons Bureau that a bomb might be made from this element, which is today called neptunium. The previous month researchers at the University of California at Berkeley had discovered that neptunium decays rapidly by beta emission to another trans-uranic element, number 94, which the Berkeley researchers named plutonium. This substance too could serve as reactor fuel or an explosive. The advantage of using plutonium rather than uranium-235 here is that it is chemically different to uranium, so separating it from uranium-238 should be much easier than separating the two isotopes. Weizsäcker did not find out about the American discoveries until after the war, but even in 1941 he understood that element 93 would decay to 94 and that this could be used in a bomb, and he drafted a patent application to that effect.

This possibility persuaded Heisenberg that an atomic bomb might not be so remote a prospect after all. Artificial trans-uranic elements, he understood, might also be prepared in a particle accelerator by bombarding uranium with protons or alpha particles. There was no such device operating in Germany during most of the war, but there was one at

Bohr's institute in Copenhagen, and construction of another had begun in Paris by Frédéric Joliot-Curie. When France was invaded, Walther Bothe and his colleague Wolfgang Genter, another member of the *Uranverein*, inspected this device and conscripted the detained Joliot-Curie to help get it running, which they did by the end of 1941.* It was used to fire a beam of deuterons – heavy-hydrogen nuclei, containing a proton and a neutron – at uranium and thorium. The reaction products were then sent for analysis to Otto Hahn in Berlin. Meanwhile, Hahn's KWIC began constructing its own accelerator in 1942: the so-called Minerva project, financed by Army Ordnance. It was never completed, but the equipment was taken to Tailfingen in South Württemberg when the institute was forced to relocate there because of bombing raids in 1944. Bothe also began constructing an accelerator in Heidelberg, which was working by the summer of 1944. While these efforts never produced any significant quantities of fissile material, they show that the German physicists understood the principles of a plutonium bomb and worked, in however preliminary a manner, towards that goal.

As the blitzkrieg war bogged down in the merciless Russian winter of 1941, Army Ordnance became more impatient to know if there was any likelihood of seeing results 'in the foreseeable future'. The physicists responded with a 144-page document arguing for the 'enormous significance' of the uranium work 'for the energy economy in general and for the *Wehrmacht* in particular'. They were walking a tightrope. If they promised more than they could deliver, they would be held accountable; but if they offered too little, they would lose their funding. The report attested that 'success can be expected shortly' on a uranium machine, and Heisenberg gave the authorities a scent of advanced weaponry without specifying how far off it might be: 'once in operation', he wrote, 'the machine can also lead to the production of an incredibly powerful explosive'. He added that if uranium-235 could be isolated (although the efforts in this direction were not making much headway), this too would constitute 'an explosive of unimaginable potency'.

---

* Joliot-Curie took advantage of his relative liberty to work for the French Resistance. When he was first seized by the Germans, they wanted to know what had happened to the heavy water that had been sent to France in 1939 by the sympathetic head of the Norwegian plant at Vemork. It had been put on a ship that sunk, Joliot-Curie told them. In fact it had already been shipped to Britain.

In February 1942, at the request of the Reich Research Council, Hahn, Harteck and Heisenberg gave lectures before high-ranking and technically literate staff representing various senior officials, including Himmler, Goering and the head of armaments Albert Speer. Speer also attended a subsequent meeting at the KWG's Harnack House in Berlin where (contrary to some reports) he seems to have been favourably impressed by the potential of the nuclear experiments. Speer himself claimed in his memoirs that the meagre funding requested by the scientists left him doubting their conviction and capabilities, but wartime documents show that in fact he took a close interest in the research, asking to be kept informed regularly about progress. All the same, the work was never granted the sort of prodigious resources made available to Wernher von Braun's rocket programme, and Army Ordnance eventually relinquished the nuclear project altogether.

At that point it became a civilian rather than military affair, for control of which the Reich Education Ministry vied with the KWG. Bernhard Rust of the REM was enthusiastic; Goering too was encouraged by the news of the research that eventually filtered through to him. Heisenberg later attested (contradicting his claim elsewhere – see page 216) that 'one can say that the first time large funds were made available in Germany was in the spring of 1942 after that meeting with Rust when we convinced him that we had absolutely definitive proof that it could be done'. 'It' here means an atomic bomb, showing that the physicists were by this stage prepared to be bold – and again contradicting Heisenberg's later suggestion that the physicists presented the bomb as at best only a very distant and abstract possibility. Heisenberg told historian David Irving* in the 1960s that 'it was from September 1941 that we saw an open road ahead of us, leading to the atomic bomb'. The promise was constantly renewed. In the spring of 1943 an official who attended a lecture by Heisenberg at the Reich Postal Ministry remembered him saying that within just one or two years the scientists should be able to deliver to the government a bomb with 'hitherto unknown explosive and destructive power'.

* Now notorious for his Holocaust denial and links to neo-Nazi organizations, Irving conducted important research on the German nuclear work in the 1960s. His books on the subject are, however, marred by too great a readiness to accept the story that Heisenberg in particular gave him, along with their sympathy – unsurprising in retrospect – for the German point of view in general.

When Army Ordnance abandoned the uranium research, the KWIP was returned to the authority of the KWG, and the occasion arose to appoint a new director to take over Debye's role formally. The acting head Diebner did not have the scientific distinction to warrant such a position, and besides he had never enjoyed the confidence of Debye's former colleagues, who wanted Heisenberg to take his place. Although Erich Schumann favoured Walther Bothe, in April 1942 the institute researchers got their way and Heisenberg was installed as director 'at' (still not 'of') the KWIP. Heisenberg suspected, perhaps rightly, that Himmler had a hand in that decision as part of his political exoneration, and wrote to him in February 1943 to say 'I thank you for the rehabilitation of my honour connected to this appointment.' The ousted and somewhat resentful Diebner went to head a rival team of researchers at Gottow, which also explored prototype reactor designs.

Not everyone was pleased with Heisenberg's appointment. Harteck considered it absurd, with some justification, that a theorist like Heisenberg should be leading such an experimentally based project. And indeed while Heisenberg understood the principles of fission (although it is debated how well, as we will see), he showed no particular flair in guiding the experiments. He divided his time between Berlin and Leipzig, where test reactors were constructed in the laboratory of physicist Robert Döpel. In Berlin Heisenberg moved on the fringes of the circle of conservative aristocrats who hatched the failed plot to assassinate Hitler in 1944. He declined an invitation to become a co-conspirator, but among those who were subsequently implicated and executed was Max Planck's son Erwin.

Despite being awarded some priority for access to materials and labour, the uranium work continued its slow pace. Experiments on the prototype reactor in Berlin didn't really commence until late 1943, by which time those Germans who were able to view the situation rationally (many were not) knew they were facing military defeat. By the end of the year conditions in Berlin, especially the heavy bombing raids, made it highly dangerous and almost impossible for research to continue. When Hahn's Kaiser Wilhelm Institute for Chemistry was almost destroyed, it was clearly time to get out. The reactor work was now conducted in a bomb-proofed basement, but that would be of little help if there was no physics institute left standing above it.

So Heisenberg began to ship the whole operation south to the Black Forest, where it was installed in a mostly vacant textiles factory in the town of Hechingen. Later, the uranium reactor itself was reconstructed in a cave in the nearby picturesque village of Haigerloch. Heisenberg moved his family to a house in Urfeld overlooking a lake in the Bavarian Alps – it looked idyllic, yet was far from that for his wife Elisabeth as she struggled to cope with food shortages and family illnesses.

Using an arrangement of uranium blocks suspended in and moderated by heavy water, the Haigerloch reactor would, in Carl von Weizsäcker's view, 'probably have become critical, in other words to have begun to deliver energy', if all the right materials had been available. 'Then Germany would have been as far [in nuclear technology] as America in late 1942', he insisted.

Getting the 'right materials' was becoming almost impossible, however. Since 1943, raids on the Norwegian hydroelectric facility had more or less dried up supplies of heavy water, although the two tons already in Germany were deemed perhaps just sufficient to get a reactor working. At the start of 1944, shortly before he was replaced by Walther Gerlach as head of the physics section of the Reich Research Council, Abraham Esau placed Paul Harteck in charge of procuring heavy water. Harteck travelled to Norway to inspect the facility, in a visit that has drawn comparison with Heisenberg's insensitive forays to occupied countries where he seemed to expect a comradely welcome from the native scientists while furthering Germany's cultural and military conquest. Harteck tried to persuade the Norwegian scientists that the heavy water was intended for pure research purposes, a claim that stretched the truth beyond recovery. And yet he berated officials in Berlin when he discovered that the Norwegians had not been paid for the heavy water they had produced, and that they were expected to bear the cost of the Allied raids. This mixture of solicitous collegiality and arrogant entitlement was characteristic of the German scientists, and perplexed and infuriated their oppressed foreign colleagues.

Problems continued to beset the heavy-water supply: in March 1944 a boat carrying a batch to Germany was sunk, apparently sabotaged. Harteck was forced to conclude that the uranium research needed production facilities inside Germany. The chemicals cartel IG Farben was asked to construct a plant, whereupon the industrial conglomerate attempted to secure patent rights – based, to Harteck's fury, on the process that he had been developing himself in Hamburg. A small-scale facility was finally built near the KWIP in Dahlem by the company Lüde in the early autumn, far too late to have any significance.

There is something surreal in this picture of Harteck pressing on with heavy-water projects, negotiating funds and industrial contracts with extortionate and truculent companies, while all around Germany is being levelled and the war is evidently entering the endgame. It is as though, so long as some kind of research appeared to be continuing, he and his colleagues could convince themselves that all was normal – that they were merely scientists doing their work under trying circumstances, making the best of a poor situation, even if that situation relied on the exploitation of occupied countries and on industrial production by slave labour.

Indeed, as Heisenberg perceived the end of the war to be imminent, he expressed the hope that he would soon be able to work untroubled by bombs or pangs of conscience, in the idyll that he had dreamt of during his days with the New Pathfinders: 'the sun will continue to shine as it has before [and] we will be able to make music and to do science, and whether or not we live richly or modestly, it will make no great difference'. He and his colleagues did not appear to anticipate any moral reckoning. Of course, they might end up being shot by the Russians or by some embittered American GI; but if they survived, they could look forward to returning at last to the lab and resuming their research, the past just a fading memory.

Prisoners of war

By March 1945 the Allied forces were advancing through Germany and dividing up the spoils. Those included the German scientists, whose knowledge of nuclear and rocket science was coveted by both the Americans and the Soviets. To this end, the Americans organized a mission to seize all the information, equipment and personnel they could find from the German nuclear research. The project was called Alsos, Greek for 'grove', a play on the name of the Manhattan Project's military director General Leslie R. Groves. Its scientific leader was Samuel Goudsmit.

Having tracked the German physicists to southern Germany, Alsos swept through the territory then under the jurisdiction of the French army, picking up Laue, Hahn, Weizsäcker and others before the French realized what was happening. But Heisenberg was the prime target, and Alsos' commander, the swashbuckling Colonel Boris Pash, raced across the 150 miles from Haigerloch to Urfeld with just a handful of troops to find the head of the uranium project sitting calmly outside his mountain house.

Heisenberg felt sure he held a strong bargaining position. Despite all the obstacles, the German scientists had come very close to achieving a working reactor, and Heisenberg and his colleagues anticipated that the Allies would be eager for their expertise. When Heisenberg was brought to him, Goudsmit found this man, who he had admired in his youth, to be supremely arrogant and apparently unaware that he was after all a prisoner. Yes, he would deign to instruct the Americans on how to build a reactor, but he couldn't possibly go there to work, Heisenberg explained, since 'Germany needs me.' Goudsmit withheld any information about the Allied nuclear programme, marvelling all the while at how the Germans were so confident that it would be inferior to theirs.

At this point Goudsmit did not yet know what had become of his own parents, who had been interned in the Netherlands and taken to a concentration camp. In late 1942 Goudsmit had asked Dirk Coster to solicit Heisenberg's help in getting them released. Heisenberg, at some personal risk, had sent a letter to Coster attesting to the good character of the Goudsmits, not knowing that they had already been sent to the gas chambers.

That was the unspoken subtext in a protracted and bitter exchange between Goudsmit and Heisenberg after the war. 'I am in a rather sad and violent correspondence with Heisenberg', Goudsmit wrote to Paul Rosbaud in 1948. 'He still does not see the bigger issues . . . All he knows about is that "his honor is being attacked" or that "German" physics is being frustrated.' 'Don't think that Heisenberg will ever agree with you', Rosbaud replied. 'He will never learn to be humble but will always be arrogant.' Whether or not that was fair, Goudsmit was himself sometimes guilty of bending the facts to besmirch Heisenberg's character and technical competence.

Only after Heisenberg's death in 1976 did Goudsmit come to any reconciliation, admitting in an obituary that 'I doubt that I or most of the physicists I know would have done better under the same circumstances.' Concerning the murder of his parents, Goudsmit realized that there was probably nothing else Heisenberg could have done. But in a way that was the whole point. These physicists who congratulated themselves on so cleverly playing power games with their leaders in the end proved to have no real power to affect the issues that mattered: who lived and who died.

11 'Heisenberg was mostly silent'

Having rounded up the German scientists, the Americans seemed unsure what to do with them. Reginald Victor Jones, a physicist in charge of intelligence for the British Air Staff, cannily offered to take them off the Americans' hands. That is how the Uranium Club came to be flown to Cambridgeshire and incarcerated in an elegant country house called Farm Hall in the little town of Godmanchester. There were ten detainees: Heisenberg, Laue, Weizsäcker and Hahn, along with Paul Harteck, Erich Bagge, Kurt Diebner, Walther Gerlach, and the KWIP researchers Horst Korsching and Karl Wirtz, specialists on the separation of isotopes. (Wirtz was also in charge of reactor construction at the Berlin institute.) Hahn and Laue had had little involvement in the wartime nuclear research, and Laue was puzzled why he was being held. But the British had their reasons for wanting him there. They figured that the presence of these senior figures would provide a moderating influence on the others, and anticipated that their own scientists would appreciate the chance to speak to them. Moreover, the Allies were already contemplating the rebuilding of science in Germany, and had identified the relatively uncompromised Laue as an ideal figurehead.*

---

* Laue's presence did provide some sober balance and realism in the physicists' discussions. But he was not immune to the prevailing sense of entitlement and affront. While the German scientists were held in Belgium awaiting a decision on what to do with them, Laue organized a weekly scientific colloquium. This routine came quickly to seem inviolable, as though the establishment of arbitrary 'traditions' might offer a veneer of academic normality. According to Harteck, when Laue was told by an English officer to prepare for transfer to England the following day, he replied 'That's impossible!' On asking why, the officer was told 'Because I have my colloquium then.' Perhaps the colloquium could be rearranged for another time, the officer suggested gently. 'But could you not have the airplane come some other time?' Laue replied.

Farm Hall was bugged with microphones connected to recording equipment, so that British intelligence might monitor the German physicists' conversations to gauge their morale and determine whether they could be trusted to cooperate in the post-war reconstruction. With characteristic overconfidence, the scientists didn't imagine that their captors would have the wherewithal for such measures. 'I don't think they know the real Gestapo methods,' said Heisenberg naïvely, 'they're a bit old fashioned in that respect.' As a result, the Farm Hall recordings are all the more historically valuable for their candour.

But this unique resource was locked away for decades after the war, first for security reasons and then because of bureaucratic complications, not to mention the objections of the surviving detainees. Samuel Goudsmit was permitted to include a few quotes from the transcripts in his 1947 book *Alsos*, but without revealing their source. The very existence of the recordings was not disclosed until the publication of Leslie Groves' somewhat self-serving memoir *Now It Can Be Told* (1962), which included some further excerpts. The full story was not really told for another thirty years, however. The transcripts were finally made public in February 1992, and were published the following year as *Operation Epsilon*, the intelligence code name for the programme. Even this account was not exhaustive – perhaps only about 10 per cent of what was recorded on shellac disks at Farm Hall was actually transcribed into the British military reports, and the disks themselves were later reused. All the same, when he first gained access to the transcripts, historian of physics Jeremy Bernstein, one of the principal documenters of the incipient nuclear age, admitted to feeling like Jean-François Champollion finding the Rosetta Stone. Now one could at last hear what the German physicists had really thought.

Private conversations

One can see why these scientists had been happy for the Farm Hall recordings to stay buried. Their tendentious, artfully constructed 'official' story of the German nuclear programme is undermined by much of what they said in unguarded ignorance of the eavesdropping. That they bicker, fret, engage in recrimination and separate into factions is surely to be expected from men of a defeated nation, worried about their families and relatives and uncertain about their future. Much more damning is the lack of any serious moral reflection on their wartime activities, or indeed on the culpability of Germany in general. They are irritated and aggrieved at being held captive, as though the victims of some grave injustice. 'Things can't go on like this', chafes Heisenberg; 'It won't do', Harteck concurs. All the same, their predicament feeds a sense of importance: 'These people have detained us firstly because they think we are dangerous; that we have really done a lot with uranium', says Weizsäcker; 'Secondly, there were important people [among the Allies] who spoke in our favour and they wanted to treat us well.' As Major T. H. Rittner, the British officer in charge of Operation Epsilon, pointed out, they did not yet seem to have quite accepted that they had lost the war.

Weizsäcker's self-aggrandizing claim that the physicists were deemed 'dangerous' because of their uranium research was deflated by the news that they heard on the BBC radio broadcast on 6 August 1945:

> The first atomic bomb has been dropped by a United States aircraft on the Japanese city of Hiroshima. President Harry S. Truman, announcing the news from the cruiser, USS *Augusta*, in the mid-Atlantic, said the device was more than 2,000 times more powerful than the largest bomb used to date . . . The president said the atomic bomb heralded the 'harnessing of the basic power of the universe'. It also marked a victory over the Germans in the race to be first to develop a weapon using atomic energy.

This was the first inkling for the German scientists that they had not been ahead of the Allies after all, but had lagged behind pitifully. Goudsmit – who had kept this information from them during the Alsos mission – confessed later that he would have dearly liked to be in the room with them that day. Hahn, who had not been directly engaged in the uranium work and so had no reputation to defend, was merciless to his colleagues. 'You're just second-raters and you might as well pack up', he said to an incredulous Heisenberg.

At first Heisenberg simply did not believe it. 'All I can suggest', he insisted, is that some dilettante in America who knows very little about it has bluffed them in saying 'If you drop this it has the equivalent of 20,000 tons of high explosive' and in reality doesn't work at all.

Even then, it appears, the German scientists were desperate to convince themselves that they enjoyed some kind of technical superiority over their Allied rivals. They clung to the belief that this gave them a strong and perhaps lucrative bargaining position, able to influence how post-war nuclear technology would evolve among the superpowers. They imagined, for example, that even if the Americans had made a bomb, they might not have got as far as the Germans in devising a controlled-fission uranium machine – a reactor. If that's so, said Heisenberg – and he convinced himself that it looked that way – 'then we are in luck: there is a possibility of making money'.

Soon enough, however, they had to accept that the Allied atomic bomb was genuine. 'I think it is dreadful of the Americans to have done it', Weizsäcker attested. 'I think it is madness on their part.' To which Heisenberg responded, 'One can't say that. One could equally well say "That's the quickest way of ending the war."' But Weizsäcker was already starting to insist on the story that he later developed with great rhetorical force. 'I believe the reason we didn't do it', he declared to his colleagues on that day of the Hiroshima announcement, 'was because all the physicists didn't want to do it, on principle. If we had all wanted Germany to win the war we would have succeeded.'

Hahn, to his credit, refused this easy evasion. 'I don't believe that', he said, 'but I'm thankful we didn't succeed.' Hahn was apparently so shaken by the news that the English guards asked Laue to make sure that he did not harm himself. Laue and Bagge kept a vigilant eye on the agitated professor late into the night, until they saw him drift off to sleep.

Confronted with the reality of nuclear destruction, the German scientists had to ask themselves if this was really what they had been pursuing in Nazi Germany. How would it seem if they were deemed to have been trying to deliver such awesome power to Hitler? And so they began to devise their exoneration. The story, as described with irony by Laue on 7 August in a letter from Farm Hall to his son in America, ran as follows:

Our entire uranium research was directed towards the creation of a uranium machine as a source of energy, first, because no one believed in the possibility of a bomb in the foreseeable future, and second, because no one of us wanted to lay such a weapon in the hands of Hitler.

Laue himself never accepted this convenient fiction. In a letter to Paul Rosbaud in 1959 he described its genesis in dismissive tones:

After that day we talked much about the conditions of an atomic explosion. Heisenberg gave a lecture on the subject in one of the colloquia which we prisoners had arranged for ourselves. Later, during the table conversation, the version was developed that the German atomic physicists really had not wanted the atomic bomb, either because it was impossible to achieve it during the expected duration of the war or because they simply did not want to have it at all. The leader in these discussions was Weizsäcker. I did not hear the mention of any ethical point of view. Heisenberg was mostly silent.

For his scepticism and general denigration of German militarism at Farm Hall, Laue told Rosbaud that he received a great deal of hostility and criticism from his fellow detainees, especially Weizsäcker and Gerlach.

Gerlach, who had replaced Abraham Esau in charge of physics at the Reich Research Council in 1944, was devastated by the whole situation, behaving rather like a defeated general. Yet Rosbaud considered him much more than a Third Reich careerist and Nazi apologist. When Rosbaud had discussed the RRC appointment with him at the time, Gerlach had readily conceded that Germany had already lost the war, and insisted that 'I don't intend to make any war physics nor to help the Nazis in all their war efforts. I just want to help physics and our physicists.' However, Rosbaud had taken exception to Gerlach's insistence on making a distinction between his country and its leaders – a fantasy shared by many intellectuals, according to which they could fight for Germany and pretend they were not at the same time supporting Hitler. Gerlach's position, according to Rosbaud, had been that 'Germany must not lose the war, but she must get rid of the Nazis.' It was a common view among the 'good' Germans: they wanted

Germany to win the war, but Hitler to lose it. 'He could not', Rosbaud later wrote,

> and probably did not want to understand that Germany, the war and Hitler could not be regarded separately, and that the war only either could be won with Hitler or lost with Hitler. I would never classify Gerlach into this group of scientists which only wanted Germany to win the war for the continuation of their own personal comfort of life and of their work. His desire was absolutely honest, he loved his country and wished the best to her and did not want her to perish.

At Farm Hall the bewildered Gerlach did not acquit himself well: dismayed that his guards did not treat him more deferentially, snapping at his colleagues, and worrying that when they returned to Germany the physicists would be held responsible for losing the war because they did not make an atomic bomb. Gerlach fretted that Niels Bohr might have helped the Americans make the bomb (in fact he did not), whereas he had vouched for the Danish physicist personally to the Nazis: it was as though he imagined they could somehow still punish him for this misjudgement. 'I went to my downfall with open eyes', he insisted, 'but I thought I would try and save German physics and German physicists, and' – the self-delusion and the aggrandizement as intransigent as ever – 'in that I succeeded.'

Myths of the bomb

The Farm Hall *Lesart*, as Bernstein calls it (the German word means a particular reading of a historical text), became the 'truth' energetically promoted by Heisenberg and Weizsäcker. To the latter it was an assertion not just of innocence but of actual moral superiority. As Weizsäcker said at Farm Hall,

> History will record that the Americans and the English made a bomb, and that at the same time the Germans, under the Hitler regime, produced a workable engine. In other words, the peaceful development of the uranium engine was made in Germany under the Hitler regime, whereas the Americans and the English developed this ghastly weapon of war.

This was the view expounded, at Weizsäcker's urging and in emotive terms, by the Austrian writer Robert Jungk in his book on the Manhattan Project *Brighter Than a Thousand Suns* (1956):

> It seems paradoxical that the German nuclear physicists, living under a saber-rattling dictatorship, obeyed the voice of conscience and attempted to prevent the construction of atom bombs, while their professional colleagues in the democracies, who had no coercion to fear, with very few exceptions concentrated their whole energies on production of the new weapon.

It is a pernicious myth. But myths, according to Mark Walker, were 'what these scientists felt they needed most'. That was, of course, no more than what many people seemed to need after the war, and in many ways still do.

Once this aspect of Jungk's book began to draw criticism, Weizsäcker and Heisenberg sought to distance themselves. Weizsäcker claimed that Jungk had exaggerated in attributing to him the story of a 'conspiracy' among the scientists to deny Hitler the bomb. Jungk in turn said that he had been misled, even 'betrayed', by the scientists. One can see his point. Soon after the book was published, for example, and before its distorted narrative had drawn much fire, Heisenberg sent Jungk a letter in which he voiced no reservations about the way it depicted the Germans' attitude to the bomb, but focused instead on burnishing the myth of how the scientists had 'resisted' the Nazis:

> I want to thank you very much for having your publisher send me a copy of your fine and interesting book about the atomic scientists . . . I find that, overall, you characterized the atmosphere among the atomic scientists very well . . . Overall, the German physicists acted in this dilemma as conservators of sort of that which was worthy and in need of conserving, and to wait out the end of the catastrophe if one was lucky enough to still be around.

The legend that the Germans intentionally avoided making a bomb for Hitler also shaped journalist Thomas Powers' 1993 book *Heisenberg's War*. Powers argued that the German physicists, and Heisenberg in particular, had not just averted the construction of a bomb because of moral scruples but had actively sabotaged it. 'Heisenberg did not simply withhold himself, stand aside, let the project die', Powers wrote. 'He killed it.' Powers adduces some wartime intelligence reports that appear to support this contention. But if there is anything substantial in it, it is very hard to see why the German physicists, so keen to exculpate themselves, failed to make that case themselves in the aftermath of the war.

Nonetheless, the 'sabotage' idea remains in currency with a few of Heisenberg's defenders.* However tenuous Powers' claims are, one must have some sympathy for him, since his case rests heavily on the apparent testament of none other than Heisenberg himself. In her 1986 book *Biography of an Idea* Ruth Nanda Anshen, Heisenberg's American editor of his memoir *Physics and Beyond*, quotes a letter that she received from him in 1970 in response to some queries that arose from a review of Goudsmit's book *Alsos*. Here Heisenberg writes that 'Dr Hahn, Dr von Laue and I falsified the mathematics in order to avoid the development of the atom bomb by German scientists.'

This extraordinary statement demands extraordinary evidence. Sadly, there seems to be none. Anshen says that the letter was among the correspondence that she gave to the Columbia University Library – but there is no sign of it in these documents today. Yet would Anshen, who admired Heisenberg, have simply made it up? Could she, aged eighty-six, be misremembering something? But then whence the full, direct quote in her book?

If we believe Anshen's account, we must then also believe that Heisenberg was prepared, if necessary, to lie outright about the German uranium work. There seems to be no alternative interpretation. The idea that the German scientists fixed the maths contradicts everything else that they, Heisenberg included, had ever said about their work towards a bomb. Neither Laue nor Hahn could be consulted about their alleged collusion in this plot, for ('conveniently', as Paul Lawrence Rose says) they were both dead by 1970. If Heisenberg was telling the truth to Anshen, he'd been spinning a lie for over two decades – and incomprehensibly so, for why would he have covered up for so long such a clear demonstration of his moral rectitude?

* It was Powers' book that first inspired Michael Frayn to write his 1998 play *Copenhagen*, about the wartime meeting of Bohr and Heisenberg, although Frayn by no means accepts uncritically Powers' account of Heisenberg's motives and character.

The affair is truly perplexing. But despite Powers' conclusions, it seems hard to see it as anything other than extremely damaging for Heisenberg. Rose believes that, when he wrote to Anshen, Heisenberg may have been troubled by the fear that his mistaken calculation about the critical mass of a bomb in 1940 was about to come to light – in which case his claim would transform an embarrassing scientific blunder into an act of heroism. In any event, Powers' version of the story is rejected by historians; his book was dismissed in the *Bulletin of the Atomic Scientists* as a fiction, while the suggestion that the Germans sabotaged the maths to deny Hitler the bomb is, in Mark Walker's view, 'tragically absurd'.

Copenhagen

Much of the debate about Heisenberg's wartime record hinges on the fact that, unlike Debye, he had much to say subsequently about his motives and goals while working under Hitler. As the war progressed, Heisenberg became one of the National Socialists' most valued ambassadors of German culture. Yet he argued later that he and his colleagues had merely bided their time under oppression, trying 'to keep order in those small corners to which our own lives [were] confined'. He recast the German physicists' meek and passive collaboration as a form of active opposition, and claimed that he had stayed in Germany purely because he wanted to help 'uncontaminated science make a comeback after the war'.

It was on a 'cultural' mission with Weizsäcker to Nazi-occupied Denmark in 1941 that Heisenberg had the meeting with Niels Bohr now made famous by Michael Frayn's play *Copenhagen*. Frayn examines the several conflicting accounts of that event, implying that we can hardly expect to arrive at absolute historical truths when we cannot even be sure of our own motivations. The historical accuracy of the play has been debated passionately – Paul Lawrence Rose lambasts it as a work of revisionism more damaging than historian David Irving's crude Holocaust denial, while Klaus Hentschel asserts that the play deserves to be admired for a 'courageous polyphony' that historians too rarely admit. But Frayn was surely right to refuse any definitive reading of the event.

What did Heisenberg expect and hope to achieve by meeting with

Bohr? Some say he was attempting to sound out his former mentor about Allied work on nuclear weapons.* Others say that Heisenberg sought Bohr's approval for his nuclear research, or his opinion on the likelihood of harnessing uranium fission. Heisenberg himself implied later that he wanted to discuss with Bohr the idea of securing international scientific cooperation over the uses of nuclear energy, and in particular to engineer an embargo on bombs. Weizsäcker insisted on this view, writing as late as 1991 that 'the true goal of the visit by Heisenberg with Bohr was . . . to discuss with Bohr whether physicists all over the world might not be able to join together in order that the bomb not be built'. This self-serving claim had been uncritically reported by Robert Jungk in 1956. David Cassidy, a leading expert on the nuclear age and on Heisenberg in particular, suggests that the possibility of making trans-uranic elements by nuclear trans-mutation had made the shadow of an atomic bomb suddenly loom alarmingly, driving Heisenberg to seek moral guidance from a father figure.

Whatever his motivation, the fact is that Heisenberg accepted an invitation to lecture on his work in astrophysics and cosmic rays during a conference at the German Cultural Institute in Copenhagen, and he travelled there with Weizsäcker, who had worked at Bohr's institute in 1933–4.

What also seems unambiguous is that Heisenberg and Weizsäcker alienated their Danish peers with their insensitivity to the hardships of occupation. As Bohr recalled, 'Heisenberg and Weizsäcker sought to explain that the attitude of the Danish people towards Germany, and that of the Danish physicists in particular, was unreasonable and indefensible since a German victory was already guaranteed and that any resistance against cooperation could only bring disaster to Denmark.' Some accounts say that Heisenberg even called the war a

---

* Arnold Kramish, for whom Heisenberg is irredeemable, makes the insinuation that he and Weizsäcker were acting as spies for the Nazis. This might seem a crude and unlikely accusation, but perhaps it depends on what one means by spying. Scientists on wartime visits like this might well be expected to provide an account of their actions, and Weizsäcker did subsequently file a report for the army saying that no work on uranium fission was being conducted in Copenhagen. 'Obviously Professor Bohr does not know that we are working on these questions,' Weizsäcker wrote, adding that 'of course, I encouraged him in this belief'.

'biological necessity'. Lise Meitner meant it as no compliment when she told Hahn after the war that '[Heisenberg's] appearance in Denmark in 1941 is unforgettable.' She was not there, but in June 1945 she wrote to Debye's former colleague Paul Scherrer describing what she'd heard about the event. This second-hand account may be somewhat exaggerated, but it offers a striking picture of the impression that Heisenberg left:

> I have heard very peculiar things about him from young Danish colleagues, about when he came to Copenhagen in 1941 together with W[eizsäcker] to stage a German physical conference and absolutely refused to see the unfairness of it. He was completely infatuated with the chimera of a German victory and set forth a theory of superior men and nations over which Germany was meant to rule.

Heisenberg sensed the hostility he aroused in Copenhagen, and was bemused by it. 'It is amazing, given that the Danes are living totally unrestricted, and are living exceptionally well', he wrote to his wife, 'how much hatred or fear has been galvanized here.' Why, they had even refused to attend his talk in the Cultural Institute, simply because the institute had previously hosted 'a number of brisk militarist speeches on the New Order in Europe'. Weizsäcker forced that issue by taking the head of the Cultural Institute, without an appointment, to see Bohr, who had no wish to be put in such a situation.

Looking back in 1948, Heisenberg recalled that their crucial discussion in Copenhagen had occurred while they were walking through the expansive Faelledpark in the centre of the city. Here he said that he had asked Bohr about the morality of working on nuclear energy, but with reference only to reactors, since he claimed still to believe that bombs would not be feasible until well after the war was over. If this was indeed what was on Heisenberg's mind at that time, it would be somewhat surprising, since there is not a single known indication that he or any other of the German physicists thought about such moral issues during that period, or indeed until confronted with Hiroshima. One rather strained interpretation is that Heisenberg hoped to signal to Bohr, without saying it explicitly, that the Germans were far away from making an atom bomb, so that Bohr might use his contacts to convey this information to the Allies and perhaps avert any attempt by them to do so. In other words, this was a covert effort to keep the world free of such weapons of destruction.

'Because I knew that Bohr was under surveillance by German political operatives', Heisenberg later told Jungk, 'I tried to keep the conversation at a level of allusions that would not immediately endanger my life.'

> The conversation probably started by me asking somewhat casually whether it were justifiable that physicists were devoting themselves to the uranium problem right now during times of war, when one had to at least consider the possibility that progress in this field might lead to very grave consequences for war technology. Bohr immediately grasped the meaning of this question as I gathered from his somewhat startled reaction. He answered, as far as I can remember, with a counter-question: 'Do you really believe one can utilize uranium fission for the construction of weapons?' I may have replied 'I *know* that this is possible in principle, but a terrific technical effort might be necessary, which one can hope, will not be realized any more in this war.' Bohr was apparently so shocked by this answer that he assumed I was trying to tell him Germany had made great progress towards manufacturing atomic weapons. In my subsequent attempt to correct this false impression I must not have wholly succeeded in winning Bohr's trust, especially because I only dared to speak in very cautious allusions (which definitely was a mistake on my part) out of fear that later on a particular choice of words could be held against me. I then asked Bohr once more whether, in view of the obvious moral concerns, it might be possible to get all physicists to agree not to attempt work on atomic bombs, since they could only be produced with a huge technical effort anyhow. But Bohr thought it would be hopeless to exert influence on the actions in the individual countries, and that it was, so to speak, the natural course in this world that the physicists were working in their countries on the production of weapons.

Could it have been a suspicion that his account would be challenged that led Heisenberg to add a mitigating comment? In any event, he qualified his statements to Jungk thus:

Everything I am writing here is in a sense an after the fact analysis of a very complicated psychological situation, where it is unlikely that every point can be accurate . . . Even now, as I am writing this conversation down, I have no good feeling, since the wording of the various statements can certainly not be accurate any more, and it would require all the fine nuances to accurately recount the actual content of the conversation in its psychological shading.

He was right to anticipate contradiction. Bohr never quite forgave Heisenberg for his conduct on that visit, although the reserved and cordial Dane did manage to resume polite social relations after the war. Yet when Bohr saw Heisenberg's account of their meeting in Jungk's book, he was so upset by its egoistical nature that he wrote an uncharacteristically angry letter that he redrafted several times but was never able to bring himself to send. Bohr was particularly incensed by Heisenberg's claim that he had dissuaded his fellow German physicists from trying to make a bomb, and that he had suggested to Bohr the idea of an international boycott on research towards nuclear weapons. He had said that Bohr seemed 'slightly frightened' by the idea that such weapons could be made – as though this had never occurred to the Danish physicist until then.

'Personally, I remember every word of our conversations, which took place on a background of extreme sorrow and tension for us here in Denmark', Bohr wrote in his unsent letter:

In particular, it made a strong impression both on Margrethe [his wife] and me, and on everyone at the institute that [you] expressed your definite conviction that Germany would win and that it was therefore quite foolish for us to maintain the hope of a different outcome of the war and to be reticent as regards all German offers of cooperation. I also remember quite clearly our conversation in my room at the institute, where in vague terms you spoke in a manner that could only give me the firm impression that, under your leadership, everything was being done in Germany to develop atomic weapons.

In another draft, he added 'you informed me that it was your conviction that the war, if it lasted sufficiently long, would be decided with atomic weapons, and [I did] not sense even the slightest hint that you and your friends were making efforts in another direction'.*

Bohr remained in Copenhagen for a perilously long time. Having a Jewish mother, he was officially 'non-Aryan', although at first the Germans were relatively lenient with the Danish Jews in order to maintain the fiction that they were in the country at the invitation of the Danish government. But in 1943 their exemption from the concentration camps was terminated, and in the early autumn the Nazis began to deport prominent Danish Jews. Bohr was tipped off about his own impending arrest at the end of September, and in early October he escaped by boat to Sweden. Fearing that he might be assassinated there by German agents, the British flew him from Stockholm to England. At the end of the year he flew to Los Alamos, where he contributed little to the technical work – 'They didn't need my help in making the atom bomb', he attested – but immensely to morale. 'He made the enterprise seem hopeful', Robert Oppenheimer, the scientific leader of the Manhattan Project, later wrote.

After the Copenhagen visit Heisenberg continued to give scientific talks throughout the territories occupied by Germany, to the satisfaction of the National Socialist leaders. According to the Dutch physicist Hendrik Casimir, during a subsequent visit to Holland in 1943 he claimed that the German domination of Europe was justified thus:

> History legitimizes Germany to rule Europe and later the world. Only a nation that rules ruthlessly can maintain itself. Democracy cannot develop sufficient energy to rule Europe.†

If it were not Germany, he warned, then Europe would be run by the Soviet Union, which would be far worse. The prospects for the

---

* These letters, publicly released after much anticipation in February 2002 in response to Frayn's play, evidently created discomfort among Heisenberg's defenders. Weizsäcker, who died in 2007, pronounced Bohr's memory 'deeply mistaken'.

† This remark was reported third-hand: it was filed in 1945 by the Dutch-American astronomer Gerard Kuiper, a member of the Alsos mission, to whom it had been related by Casimir. Casimir alleged that Heisenberg had said this to him during a private walk in Leiden, during which he had also learnt that Heisenberg knew at that time about the German concentration camps. It isn't clear that Casimir's report can be taken at face value.

German war were by then looking very dim, but Heisenberg's own prestige was never greater.

While there is now no definitive way to discover what was really said in Copenhagen, we can certainly make inferences about Heisenberg's role in wartime German science. He was in Copenhagen on behalf of the conquering power, and he anticipated that the Danish scientists would be reassured in seeing it represented by a friendly face. In other words, he felt that his personal status would somehow negate all the indignities of occupation: the same kind of grandiosity that left Heisenberg convinced that he must remain in Germany come what may, since only he could rebuild German science after the war. He had become lost in self-regard.

Saving face

Why didn't the Germans build an atomic bomb? Would they have done so if they could?

After the war, Heisenberg was determined to show that he and his colleagues had been in command of their situation and had engineered its outcome. 'From the very beginning', he wrote in a description of German nuclear research published in *Nature* in 1947 (a translation of a piece that first appeared the previous December in *Naturwissenschaften*), 'German physicists had consciously striven to keep control of the project.' This, he said, was made possible by the convenient fact that making a bomb was neither impossible (in which case the question of whether they should would not have arisen) nor easy (in which case they certainly couldn't have prevented it):

> The actual givens of the situation, however, gave the physicists at that moment in time a decisive amount of influence over the subsequent events, since they had good arguments of their administrations – atomic bombs probably would not come into play in the course of the war, or else that using every conceivable effort it might yet be possible to bring them into play.

In 1968 Heisenberg offered a somewhat less triumphalist version of this story:

Obviously we were not *fully* aware of the extent of the danger, but then in the first two years the following became evident: it was possible to build nuclear reactors relatively easily with moderate means, i.e. certainly in the period of a few years available to us, that is, *that* particular reactor we knew would work, namely a reactor using natural uranium and heavy water. It was *also* clear that an explosive is produced in such a reactor [plutonium] from which atom bombs could be made; but luckily it was *also* clear then that this would involve an enormous technological investment lasting many years. Therefore we could report these results with complete honesty and a good conscience to the government agencies, and the consequence was – just as we had hoped – that the government decided to make *no* effort to construct atom bombs, but that we received certain – albeit modest funds [*sic*] – to continue work on the design of a reactor, precisely on a heavy-water reactor.

This is a carefully constructed tale. It emphasizes that the physicists were totally competent: they knew the reactor would work, they knew bombs could be made from it, they knew that would be extremely difficult (although, given the relative ease of separating plutonium, Heisenberg overplays this technical challenge). They provided the Nazi leaders with honest information, albeit tailored to secure continued funding without committing them to a bomb. Moreover, by stressing the modesty of those funds, Heisenberg can explain why they didn't get as far as the Manhattan Project. This judicious management of the situation meant that, as he put it in *Nature*, they were 'spared the decision' of whether to work on a bomb for Hitler. Such statements by Heisenberg after the war made no mention of his explicit promise to the authorities in 1942 that a 'uranium machine' could produce a powerful explosive.

Thus Debye's half-humorous account to Warren Weaver of how the German physicists considered that they were duping the authorities into funding fundamental research was, for Heisenberg, the full reality of the matter:

The official slogan of the government was 'We must make use of physics for warfare.' We turned it around for our slogan: 'We must make use of warfare for physics.'

Ours was the noble goal of knowledge, Heisenberg is implying, which we pursued by a clever ruse. The story places the scientists at the helm, reduces their political leaders to a bunch of credulous fools, makes their state-funded military research almost an act of resistance, and separates pure, untainted science from the nasty realities of the Nazi regime.

But how much of it can we believe?

The suggestion by Heisenberg and his colleagues that they purposely slowed the pace of research to keep a terrible weapon out of Hitler's hands was furiously disputed by Samuel Goudsmit. He insisted that the scientists would certainly have made a bomb if they had been able, but that they were prevented from doing so both because of the incompetence of their political management and because they themselves did not understand how to do it. He wanted Heisenberg to acknowledge what Nazi rule had cost German science, and thus to show how science can flourish only in a free society. But he also wished to see some recognition in the German physicists of their own arrogance and complacency, thinking that they alone could solve the problems of harnessing nuclear fission. Heisenberg, however, refused to accept that the best German scientists – not Nazi pawns like Diebner – were anything other than highly competent. The two men entered into protracted and sometimes intemperate correspond-ence, but neither seemed willing to recognize the real obstacle to the German bomb: that the Nazis were never sufficiently persuaded of its feasibility to allocate resources on the scale of the Manhattan Project. Given that, Heisenberg was right to say that they were spared the ultimate moral decision. All the same, Goudsmit put his finger on the crucial point, which applied not just to Heisenberg but to almost all of the German physicists: he 'fought the Nazis not because they were bad, but because they were bad for Germany, or at least for German science'.

Heisenberg was stung by Goudsmit's imputation that the Germans hadn't made a bomb because they never figured out how – an idea promoted also by Leslie Groves in *Now It Can Be Told*. Their sense of intellectual superiority had suffered badly from the discovery that the Allies had been so far ahead of them, and while they insisted that the reason was solely the level of funding provided by the respective state leaders, that suspicion remained: had the Germans not argued more strongly for support because they were mistaken about the magnitude of the task?

A key aspect of this question relates to the amount of material needed to make a bomb: the critical mass above which a fissile substance would develop a spontaneous, runaway chain reaction. In the report to Army Ordnance in 1942 the critical mass of uranium-235 or plutonium was estimated as 'presumably around 10–100 kilograms'. Yet even this vague figure seems not to capture the extent of the uncertainty. The estimates were probably made by Heisenberg, whose statements on the matter give a very confusing picture. When he heard about the Hiroshima bomb at Farm Hall, Heisenberg reacted with disbelief because he could not believe that the Allies could have produced several tons of pure uranium-235. To this, Hahn responded, 'I thought that one needed only very little "235" . . . if they have, let us say, 30 kilograms of pure "235", couldn't they make a bomb with it?' Heisenberg replied that 'it still wouldn't go off': he didn't believe it was enough. 'This statement', comments Jeremy Bernstein, 'shows that at this point Heisenberg has no idea how a bomb works.'

If Heisenberg had forgotten his earlier estimates of the critical mass, Hahn had not. 'But tell me why you used to tell me that one needed 50 kilograms of "235" in order to do anything', he demanded. 'Now you say one needs two tons.' Heisenberg was evidently disconcerted, and he dissembled: 'I wouldn't like to commit myself for the moment.' He went away and made a better calculation, deciding that indeed only a few tens of kilograms would suffice.

Heisenberg later claimed that he had long known the critical mass to be quite small. In 1948 he told Samuel Goudsmit that in his meeting with Albert Speer and other officials in Berlin in June 1942 he was asked how large a bomb would have to be to destroy a city, and answered 'about the size of a pineapple'. 'This statement', he stressed to Goudsmit, 'of course caused a surprise especially with the known physicists and it has therefore remained in the memory of several participants.' This was the only time Heisenberg claimed to have a definite and accurate estimate of the critical mass before his calculation at Farm Hall, and it's not clear what reasoning lay behind it.

Paul Lawrence Rose argues that Heisenberg in fact made a severe overestimation of the critical mass in 1940, based on a misunderstanding of the physics involved, and that only after challenged by

Hahn at Farm Hall did he work through the theory properly. Rose believes that Heisenberg subsequently evaded the issue because he did not want to admit his mistake, and that Heisenberg's post-war story intentionally obscured the truth that he and his colleagues would have developed a bomb had they not been deterred by this misapprehension that the critical mass was unattainably immense.

Rose asserts that a failure to distinguish the physics of bombs from that of reactors (which cannot generate a nuclear explosion, although an uncontrolled chain reaction can lead to excessive heat production and meltdown) led the Germans to believe that one could make a kind of hybrid 'reactor-bomb'. Indeed Heisenberg speculated, on hearing the news of Hiroshima, that perhaps the Allies had dropped such a chimeric device. And according to one version of the Copenhagen encounter, Heisenberg gave Bohr a drawing of the German bomb design when they met, yet when Bohr passed this on to the scientists at Los Alamos they thought it looked more like a reactor than a bomb. This story was told to Bernstein by Hans Bethe, who worked at Los Alamos. But Bohr's son Aage has always strongly denied that Heisenberg gave his father such a sketch, and indeed it seems rather unlikely given the caution that Heisenberg displayed about disclosing information in their meeting. Nonetheless, the ambiguous bomb/reactor drawing does seem to have turned up at Los Alamos from some source or other.

Other historians think Rose has exaggerated the physicists' theoretical shortcomings. Rainer Karlsch and Mark Walker say that the patent application filed by Weizsäcker in 1941 for making a plutonium bomb 'makes it crystal clear that he did indeed understand both the properties and military applications of plutonium'. Walker says that 'as far as they got, [the German scientists'] understanding was comparable to what the Americans and émigrés did'. He considers the accusation that the German physicists lacked technical competence to be another of the myths told about the German bomb.

Be that as it may, faced with the evidence from Hiroshima that a bomb *could* have been made within just a few years, Heisenberg and colleagues needed to explain why they had let their leaders believe otherwise, without having to admit that this was because of any technical error. In his 1947 *Nature* paper Heisenberg placed the onus for the funding decision on the Nazi government. He argued that the scientists had known perfectly well that a bomb could have been made with uranium-235 or plutonium, and that their research was on a par with that in the United States until 1942, when Albert Speer allegedly decided to withdraw most of the support and put it instead into the rocket programme. This, Heisenberg claimed, suited them fine: they could get on with research on the peaceful uses of nuclear energy that would be needed once the war was over:

> We could feel satisfied with the hope that the important technical developments, with a peacetime application, which must eventually grow out of [Hahn's and Strassmann's] discovery, would find their beginning in Germany, and in due course bear fruit there.

That is far too tidy a picture. For one thing, if the German scientists knew as much as the Allies and were on the same footing until 1942, how come Enrico Fermi's reactor in Chicago went critical that year while the Germans never succeeded even in building a working reactor during the entire war? The Germans pleaded disruption by bombing raids, and lack of adequate government support. But the latter, at least, does not match the fact that nuclear research enjoyed a rather high priority for resources – greater, for instance, than aircraft production. In the 1950s Speer, under arrest for war crimes, said that he hoped Heisenberg would not try to place the blame for the failure of the uranium work on *him*. But evidently that had already happened.

Another false trail in the 'official' German story is the question of the neutron moderator. Walther Bothe was made a scapegoat for the failure of the reactor project because his 'erroneous' measurements of neutron absorption in graphite had excluded it from consideration in favour of heavy water, shackling the efforts to the travails of heavy-water production. Heisenberg in particular laid the blame on Bothe for why the uranium research was retarded. This was not only unjust but untrue. While Bothe's studies indeed seemed to indicate that graphite wouldn't work, Wilhelm Hanle found subsequently that the excessive absorption seen by Bothe was caused by impurities in the material – very pure graphite should work just fine. However, the cost of such purification was deemed by Army Ordnance to be too high.

In the end, the relevance of technical matters such as the critical mass to the question of why the Germans did not make a bomb is unclear. Their lack of progress in separating uranium isotopes, and the fact that they never quite managed to make a uranium reactor that could sustain a controlled fission chain reaction, meant that even a very modest estimate for the critical mass would have seemed unattainable during the war. It is not clear that the funding decisions of the German authorities ever hinged on fine details of reactor or bomb design and engineering, about which even the scientists seemed very vague. Rose's suggestion that all depended on what Heisenberg did or didn't know ironically echoes the physicist's own grandiose belief that the uranium project, if not all German physics, depended on him alone.

In the post-war years, Heisenberg and Weizsäcker oscillated between suggesting that they were passively 'spared the decision' of whether to make a bomb because of a lack of funding, and that they actively manipulated the situation so that there was no prospect of them ever having to face the dilemma. Weizsäcker even claimed in 1993 that he had participated in the research hoping that those with the technical knowledge of such an awesome weapon would become so indispensable to the Nazis that they might be able to influence Hitler's policies. Did he truly think he might, then, prevail on Hitler to close the concentration camps? Weizsäcker himself seemed to sense how implausible this sounds, stressing that it was a 'dreamy wish' if not indeed a little crazy. In any case, how did he think such a motive would carry force unless the physicists had been able to demonstrate that they could fulfil their promise and liberate nuclear energy?

But it never seemed greatly to matter to Heisenberg and Weizsäcker whether these stories were rigorous, consistent or plausible. It was enough that they should confer some degree of moral impunity. The evidence now excavated from the war years undermines these fictions. So were Heisenberg and Weizsäcker sincere but self-deceiving, or actively attempting to mislead? Paul Lawrence Rose, inclined to believe the very worst of these men at all times, considers their stories a fantasy concocted to preserve their dignity, reputation and 'honour' – the latter being understood in the distinctly German sense of one's inner integrity, rather than (as others might see it) the moral orientation of one's actions. Mark Walker, on the other hand, argues that it was not so much, or not entirely, self-interest that shaped their accounts, nor even fear of being denounced as Nazi stooges, but the fervent wish to preserve the reputation of German science. There seems in any event to be more behind them than a selfish attempt to appear blameless and untainted by Nazi corruption.

Can we, in the end, say that the German scientists tried to make a bomb or not? As Walker has argued, it isn't a good question, precisely because it sounds as though it should have a simple answer. The words are deceptively ingenuous. What, for example, do we mean by 'tried'? The physicists knew it should be possible, and they undertook the initial stages of the programme, such as developing techniques for separating isotopes. They were aware that if a 'uranium machine' could be made to work then it would produce at least one new fissile element. 'Explosives' featured repeatedly, if not ubiquitously, in their appeals to the Nazi leaders. But neither the scientists nor their leaders regarded an atomic bomb as a significant priority, for none of them believed it could be done in short order. The physicists did not argue a strong case for creating a bomb because they lacked the conviction that they could achieve it in the near term.* A misunderstanding of the physics involved might have played a part in that, but it was probably not the determining factor. The German government did not lack the funds – the Peenemünde rocket programme cost a comparable amount to the Manhattan Project – but mercifully they, too, had insufficient faith.

Documents confiscated from the KWIP by the Russians and recently returned to the Max Planck Society hint that in fact the Germans *did* build an atomic bomb, after a fashion. These papers seem to indicate that Kurt Diebner's group at Gottow produced two small 'nuclear' explosions in Thuringia, eastern Germany, in March 1945, killing hundreds of prisoners of war and concentration-camp inmates conscripted as slave labourers. Let alone anything else, if this dramatic claim is true then it implicates those physicists directly in war crimes.

It is perfectly possible that non-nuclear explosions occurred during the uranium research, especially because water may react with uranium to produce flammable hydrogen gas. In June 1942 one of

* Towards the end of the war, the scientists made some suggestions that a bomb might be only a year or two away – but even then, no extensive, industrial-scale effort was engaged to make it happen, not least because bombing and competing war priorities rendered that impossible.

Heisenberg's prototype reactors at Leipzig, developed with Robert Döpel, was destroyed in a hydrogen explosion from which they were both lucky to escape unscathed. But the Gottow detonation seems to have been no accident. The accounts point to an attempt to trigger either fission of isotope-enriched uranium or, still more extraordinarily, nuclear fusion of deuterium, with a blast using conventional explosives to set off a concentric spherical device with a uranium or plutonium core. It's certainly the kind of wild, desperate act that the last days of the war provoked. If it happened, Heisenberg and Weizsäcker seem to have known nothing about it, although allegedly Walther Gerlach was aware of and approved the tests. Yet even Diebner appears not to have believed that there had been any precursor to the Hiroshima bomb: at Farm Hall he exclaimed that 'We always thought we would need two years for one bomb.'

Whitewash

In 1946 Heisenberg was permitted to return to a shattered Germany, where he settled in Göttingen and established the Max Planck Institute for Physics – in effect a reconstituted KWIP, now officially bearing the name that Debye had chosen. Göttingen was designated by the Allies as the hub of science reconstruction in what would soon become West Germany. Weizsäcker was made director of theoretical physics at Heisenberg's institute. He became a pacifist, campaigned for nuclear disarmament, and founded a centre for 'science and peace research' in Hamburg. In 1984 his brother Richard became president of the Federal Republic of Germany and presided over the reunification. In 2009, two years after Carl's death, the National Academy of Sciences Leopoldina, Germany's oldest scientific society, inaugurated the Carl Friedrich von Weizsäcker Award for 'scientific contributions to socially critical questions'. Despite this sincere concern for science's social role, on his wartime activities Weizsäcker never clearly and unequivocally voiced a word of regret.

In September 1946 the Kaiser-Wilhelm-Gesellschaft, having been briefly dissolved, was resurrected as the Max-Planck-Gesellschaft.*

* Until 1949 this institution was recognized only in the British zone of occupied Germany.

Planck's name, it was thought, would expunge any association with the Nazi regime, an aspiration reinforced by the appointment of Laue as president. Whether Planck himself would have accepted that symbolic role, we do not know. He too settled in Göttingen, not only elderly now but broken in spirit. He had suffered terrible personal tragedies: having lost his first son in the First World War, the life of his second had been taken by the Nazis following the plot to assassinate Hitler. Both of his daughters had died in childbirth before Hitler came to power.

Planck died in 1947. Posterity recognizes his goodness, but it has become increasingly clear how inadequate that alone was for navigating the challenges Planck faced. Historian Dieter Hoffmann, who is seldom sparing in his criticisms of the German physicists of that era, said on the fiftieth anniversary of Planck's death that 'he stands for professional excellence and the sustained search for truth, for scientific and personal integrity, for humanity and truthfulness, humility and modesty'. But to his biographer John Heilbron, Planck was a man catastrophically and tragically betrayed by his own beliefs:

> What he did during the Nazi period was to act in accordance with a world view that allowed him no escape from his situation with his honor intact.

The rebuilding of German physics began by salvaging its reputation. That demanded more myths. In 1946 the vice chairman of the DPG, Wolfgang Finkelnburg, wrote in the society's journal *Physikalische Blätter* that

> physicists have a right to know how, despite all of the difficulties and with much courage, the Executive Committee of the German Physical Society did everything in its power in the years after the last physicists' conference in 1940 to represent to the [National Socialist] party and the [Reich Education] ministry a clean and decent scientific physics and to prevent worse events than already occurred. I believe that this fight against party physics may be regarded as a heroic chapter of the real German physics, because – although led actively by only a few – it was effectively and morally supported by the predominant majority of physicists.

When we are compromised by our actions, the urge to evade blame and to construct a story that we can live with is one that everyone can recognize. Yet Dieter Hoffmann is harsh but not unfair when he says that the DPG's chairman Carl Ramsauer and his colleagues contributed to the broader conspiracy of silence in post-war Germany about the realities of the Third Reich. There was no 'heroism' in the DPG's activities, he says, but only – at best – damage limitation. The 'formula of exoneration' applied by Ramsauer and others, Hoffmann says – the notion 'that they had done everything possible for science, and implicitly only for noble causes' – leaves out entirely 'any consideration that they had conducted their science within and for a criminal regime that they had supported and worked for to gain personal and professional advantages'.

Finkelnburg's statement illustrates how the *Deutsche Physik* movement (now labelled 'party physics') was exploited retrospectively as a way to distinguish the 'clean and decent' majority of physicists from the Nazis. In effect it served as a receptacle for any taint of collaboration with the regime. Finkelnburg does not mention that after 1940 *Deutsche Physik* was all but finished anyway, nor that this was never the 'official' physics of the National Socialist leadership, who had tended to regard the physicists' battles with bemusement. No, what German physics now needed were scapegoats. That was true throughout German society, of course: it became common practice to load collective guilt for Hitler's regime on to a few 'true Nazis'. But in physics this went one stage further: in the narratives presented by Heisenberg and others, those scientists who had most enthusiastically supported the National Socialist agenda were also the least competent. Thus German science could be redeemed not only politically but also professionally.

This fractionation of scientists into 'Nazis' and 'non-Nazis' can be witnessed even at Farm Hall, where the physicists began to turn on each other. It was obviously compromising now to have been a party member, as Bagge and Diebner were. Both tried to excuse themselves by arguing that they had never held real sympathy for the National Socialists. Diebner said that this membership was purely a matter of expedience – he'd hoped it would improve his job prospects once Germany had won the war – and he listed his various acts of 'opposition'. His colleagues weren't impressed; some said they could not in good conscience sign a declaration of their anti-Nazi stance during the war if Diebner was also a signatory. The younger Bagge claimed rather pathetically that his mother had enlisted him in the National Socialists without his knowledge, a task impossible even for the most Machiavellian of mothers.

The issue was something of a red herring in any case. With the exception of Laue, Heisenberg and Hahn, all the others had belonged to a National Socialist organization, and Heisenberg had given his services to the government willingly enough. Major Rittner witnessed such nationalistic chauvinism amongst the detainees that he felt compelled to allude in a report in September to 'the inborn conceit of these people, who still believe in the *Herrenvolk*'. Laue, he said, is the only possible exception.

Very few Germans in any walk of life suffered repercussions as a result of being alleged Nazi sympathizers. Many of the scientists – the physicists especially – did have genuinely valuable intellectual wares to trade for their rehabilitation, and this sort of barter was acceptable to both the Americans and the Soviets (Robert Döpel at Leipzig was one of the nuclear physicists who went east). Even the 'Aryan physicists' Stark and Lenard escaped serious recrimination. They were called before the denazification tribunal, but Lenard was deemed too frail to stand trial. Stark was at first classified as a 'major offender', the most serious category of the five-point ranking, and was sentenced to six years of forced labour. But an appeals court commuted the classification to 'lesser offender' (the third category), and ordered Stark simply to pay a 1,000-mark fine.

That was a typical pattern. The post-war trials were notoriously ineffectual, since it was extremely difficult and time-consuming to investigate any allegation thoroughly, let alone to prove it. Literally millions of cases were simply dropped. Many who supported the regime had little difficulty in obtaining the so-called *Persilscheine* or whitewash certificates. The most vociferous of Nazis in the universities were dismissed without compensation, while others who had doubtless helped the regime were eased into early retirement. Pascual Jordan, for example, a party member whose enthusiasm for National Socialism was such that its ideology has been said even to have seeped into his physics, was issued a whitewash certificate by Heisenberg, who attested that he had 'never reckoned with the possibility that

[Jordan] could be a [true] National Socialist' (rather inviting the question of what it would take to convince Heisenberg of that).*

This situation left Paul Rosbaud thoroughly disillusioned. In a letter to Samuel Goudsmit in 1948 he wrote that

> Most of our old friends are either back in their jobs or at least denazi-
> fied or busy to get testimonials – and they will get their testimonials
> . . . They will show you some nice letters from people whose names
> I don't want to tell you and you will learn from these letters that they
> have been very nice fellows and sometimes have even said nasty things
> about Adolf.

The denazification of German science was actively obstructed even by those who had had no sympathy for the National Socialists. The prevailing attitude was one of resentment at the intrusions of the occupying Allied authorities, which led to a closing of ranks and a feeling of solidarity between the most unlikely of bedfellows. Even relatively blameless individuals refused to condemn those who had been clearly implicated in the Nazi regime. Walther Gerlach, for example, issued a *Persilschein* for the SS officer Rudolf Mentzel, with whom he had by no means seen eye to eye during the war.† And Laue and Sommerfeld supported efforts to lighten the sentence meted out to Stark at Nuremberg – an expression not, it seems, of saintly forgiveness but of professional allegiance.

Others drew an invidious parallel between the purging of Nazis after the war and the persecution of 'non-Aryans' before it. Faced with accusations against the unambiguously pro-Nazi Pascual Jordan and Herbert Stuart, Otto Hahn complained that 'we had enough trouble with all that snooping and telling off during the Third Reich'. For Hahn, denazification involved 'attacks against the science of our

---

* Niels Bohr was less obliging: he replied to Jordan's request for a letter of exoneration by sending the physicist a list of Bohr's friends and relatives who had died under the Nazis.
† In the trials Mentzel was identified as a 'lesser offender' and sentenced to two and a half years in prison. But he was immediately released after the sentence was pronounced in early 1948 because his internment since the end of the war was deemed to have satisfied it already. His boss Bernhard Rust would probably have fared little worse, had he not committed suicide in May 1945.

nation' – once again it seems he felt that the 'integrity' of German science must be defended at all costs and could be detached from the political agenda of the Third Reich.

These prevarications and evasions during denazification meant that it quickly became impossible to construct a clear picture of how the Nazification of German society had proceeded in the first place. 'It was one of the most depressing experiences I ever had as a historian', says Klaus Hentschel, 'to see reflected in the documents how very soon *after* 1945 the chance of coming to grips with the National Socialist regime was allowed to slip away, thus missing the opportunity to make a frank assessment of the facilitating conditions the regime had set.'

This refusal to address wartime conduct continued to frustrate scientists outside Germany for many years. It did not seem to matter what one had done, so long as one could say (often quite truthfully) 'I never liked the Nazis.' The prevailing attitude was not guilt or remorse but self-pity and resentment at the indignities now being inflicted. Visiting Germany in 1947, Richard Courant, the mathematician who had been forced out of Göttingen in 1933, despairingly described its residents as 'absolutely bitter, negative, accusing, discouraged and aggressive'. Hartmut Paul Kallmann, post-war director of the former KWI for Physical Chemistry in Berlin, who as a 'non-Aryan' had been dismissed under Haber's directorship in 1933 and had worked for IG Farben during the war, wrote to the émigré Michael Polányi in 1946 saying that 'the tough momentary situation [here] is deplored much more than the evil of the past 10 years . . . The masses still don't know what a salvation the destruction of the Nazis was to the whole world and to Germany as well.' 'It is a difficult problem with the Germans', Margrethe Bohr told Lise Meitner two years later, 'very difficult to come to a deep understanding with them, as they are always first of all sorry for themselves.'

Sometimes it was worse than that, for one should not imagine that all Germans felt an urgent need to distance themselves from the Nazis. In 1947 the president of the polytechnic at Darmstadt complained that for some students 'it seemed that the only thing the Nazis had done wrong was to lose the war'. Kallmann eventually quit Germany in 1949, giving up his positions at the chemistry institute and the newly created Technical University in Berlin because he was disgusted at the Nazi mentality that, he felt, still dominated academia.

Heisenberg exemplifies this denial of the past. He was apt to refer to 'the bad side of Nazism', with the implication that there was a 'good' side too. He seemed, even after the revelations of Auschwitz, to remain stubbornly blind to the character of his leaders, insisting that if Germany had won the war then in time – he gave it fifty years – the Nazis would have become civilized. 'He still goes on defending all the evil things in Germany as being the normal by-products of any social revolution', wrote Goudsmit to Rosbaud in 1950 after meeting Heisenberg in the United States. That Heisenberg could peddle this naïvely optimistic line in 1947 even to one refugee physicist in Britain who had lost his job and then friends and relatives in the concentration camps makes it clear that there is something to be explained in Heisenberg's character which accusations of ambition and arrogance don't quite account for.

An aversion to self-examination has been disturbingly long-lived in German science. In the KWG's institutes for anthropology, medical sciences and psychiatry there was far graver accommodation and active collaboration – sometimes with horrific consequences – than one finds in the compromises and prevarications of the physicists: for example, Otmar von Verschuer, director of the KWI for Anthropology, Human Heredity and Eugenics considered Joseph Mengele to be his collaborator.* This ugly legacy has been well documented, yet even in the 1980s the Max Planck Society was reluctant to face up to it. When, after becoming MPG president in 1997, the biologist Hubert Markl bravely commissioned a project to investigate the society's role in Nazi Germany, there were grumbles that the programme would foul the society's own nest. Only in 2001 was the MPG ready to make a public apology and admission of guilt and complicity for the criminal medical research of Mengele and his ilk. At a conference to which the few survivors of the atrocities were invited, Markl said

I would like to apologize for the suffering of the victims of these crimes – the dead as well as the survivors – done in the name of science . . . when I apologize here personally and for the Max Planck Society

* Verschuer remained a professor of genetics at Münster until 1965.

representing the Kaiser Wilhelm Society, I mean the honestly felt expression of the deepest regret, compassion, and shame over the fact that scientists perpetrated, supported, and did not hinder such crimes.

Paul Lawrence Rose argues that there is a trait specific to German culture that led to an inability – not simply a 'refusal' – on the part of the scientists and other intellectuals to appraise their behaviour under the National Socialists in the kind of moral terms that other Western nations might have expected. Rose even goes so far as to treat the 'German' mentality as distinct from what he calls 'Western'. He adduces a centuries-old tradition in Germany of equating morality with individual autonomy of thought (*Innerlichkeit*), not with external actions such as resisting political evil. Faced with a corrupt state, this tradition required that one seek only to preserve some 'inner freedom', while permitting and even demanding complete obedience to the rulers. It is in this sense, says Rose, that the population of Wilhelmite Germany could believe themselves to be free even under an authoritarian monarchy, while those who went into voluntary exile from the Nazis, such as Thomas Mann, were widely held in contempt by those who remained, even if they disliked the Nazis, for treasonous dereliction of 'German culture'.

Although the assertion of a 'German mentality' seems troublingly close to that of a 'Jewish mentality', Rose's analysis does seem to fit with the attitudes of Heisenberg and Weizsäcker, whose sometimes perverse statements and actions during and after the war cannot plausibly be ascribed to latent Nazi sympathies. 'The conditioning of German culture and behavioral patterns made the mentality and feeling of Heisenberg and his colleagues an alien intellectual and moral universe that their Allied counterparts could only regard with disgust and bemusement', Rose asserts. Many Germans today will attest to this attitude among the generations brought up in the first half of the twentieth century.

While it seems simplistic to lay all the moral myopia at the doorstep of pre-war 'German culture' – it will not get us far in understanding the diversity of responses from the likes of Laue, Schrödinger, Planck, Rosbaud and Debye – it does appear that the self-justification of German scientists after the war was not so much an act of evasion as a genuine belief that there was nothing to feel guilty about.

And to outsiders, this attitude did and does remain nigh incomprehensible. To feel no responsibility at having worked under, and in some sense *for*, a racist, genocidal gang of criminals seemed to indicate a sheer absence of moral reason. That clash of values can be discerned in the lengthy exchange of letters between Lise Meitner and the astrophysicist Walter Grotrian, who were friends in Berlin before Meitner fled. Wishing to resume relations after the war, Meitner wrote to Grotrian to say that she needed to know from him how he could have reconciled himself to visiting an observatory in Tromsø in occupied Norway, as though nothing were amiss in his dropping by on behalf of his Nazi leaders – an action in many ways comparable to Heisenberg's and Weizsäcker's trip to Copenhagen in 1941. Grotrian seemed merely baffled by her complaint, so Meitner spelt it out for him: 'it remains incomprehensible to me that a fair-minded scientist – and that is what I have always known you to be and valued you for – would consider it an appropriate mission to organize scientific work in an unlawfully occupied country for the benefit of those in power'. Grotrian replied that he had at that time accepted the 'official' reasons for why the Germans had invaded Norway, even if they were later shown to be fallacious, and that his visit had been purely scientific. 'With your completely different kind of attitude', he wrote, 'you are unlikely to understand my way of acting.' Meitner could not reconcile why such a basically decent man could have ever, by 1940, considered the Nazis to be leaders with whom one could and should work. Grotrian, who had even enlisted in the Luftwaffe, failed to see where the problem lay.

Many scientists outside Germany felt by the late 1940s that their German counterparts had an easy ride. Some in the United States were particularly dismayed to see Germans, unabashed at their wartime research, being granted special dispensation to enter the country and work for the American government, most notoriously the architect of the flying bombs Wernher von Braun.* 'It is, in most cases, morally wrong for our scientists to collaborate with these imported colleagues', wrote Goudsmit. 'Those who opposed the

---

* The V-1 and V-2 rockets killed around 15,000 people in Britain and Belgium; but in some ways it is yet more appalling how many perished in their manufacture: around 20,000 slave labourers died in Peenemünde, having worked under unthinkable conditions.

excesses of the Nazi regime, were nevertheless in agreement with its policy of an imperialistic Germany, ruling the world. I know of only very, very few who clearly saw the German errors and acted accordingly.'

'Armorers of the Nazis'?

It was partly to counter the distortions propagated by the German physicists that Goudsmit wrote his 1947 book on the Alsos mission. In a review of that book, the Manhattan Project physicist Philip Morrison wrote:

> The documents cited in *Alsos* prove amply, that no different from their Allied counterparts, the German scientists worked for the military as best their circumstances allowed. But the difference, which it will never be possible to forgive, is that they worked for the cause of Himmler and Auschwitz, for the burners of books and the takers of hostages. The community of science will be long delayed in welcoming the armorers of the Nazis, even if their work was not successful.

Laue felt stung into reply, notwithstanding Morrison's remark that 'brave and good men like Laue could resist the Nazis even in the sphere of science'. Rightly regarded outside Germany as almost unique among German scientists in the integrity of his resistance to Hitler's rule, Laue commanded an unparalleled degree of respect and moral authority. Yet his behaviour in the immediate post-war period is more ambivalent, displaying unseemly haste to lay the past to rest in the interests of his profession and his nation. What he offered to Morrison's charge was an apologia – honest, sincere and in many ways commendable, but an apologia nonetheless:

> Ever since war between civilized states relapsed once more into the old barbaric 'total' war between people, it has been no easy matter for an isolated citizen of a warring nation to withdraw himself altogether from war service . . . If one or other among the German scientists found it possible during the war to avoid being drawn with his work into the maelstrom, it is not allowable to conclude that it was so for all.

The directors of the larger research institutes in particular were under the absolute necessity of putting the facilities of their institutes at least partially and formally at the service of the war effort. Open refusal on their part, immediately classable as 'sabotage', would have led inexorably to catastrophic consequences for themselves. On the other hand, an (often fictitious) compliance with the demands of the armed forces had advantages which our opponents should recognize as legitimate.

Laue argued that most of the science done in Germany during the war was 'honest, solid scientific investigation, following steadily in the steps of the preceding peacetime research' and had 'nothing whatso-ever to do with Himmler and Auschwitz'. Besides, he said, while recognizing 'what unutterable pain the mere word Auschwitz must always evoke' in Goudsmit, this must surely make him incapable of unbiased evaluation. 'How careful one must be', Laue wrote, 'in passing judgment on events which took place under a tyranny.' One cannot help being moved by his final appeal not to let the bitterness and hate linger:

> We recommend as the foundation of every utterance of peace politics, in great and small things alike, the words which Sophocles puts in the mouth of Antigone, citizeness of a victorious state: 'To league with love not hatred was I born.'

Yet Morrison, a scientist of uncommon wisdom and humanity, was not dissuaded. 'Many of the most able and distinguished men of German science', he replied,

> moved doubtless by sentiments of national loyalty, by traditional response to the authority over them, and by simple fear, worked for the advantage of the Nazi state. These men *were* in fact the armorers of the Nazis. Professor Laue, as the world knows and admires, was not among them. It is not for the reviewer to judge how great was their peril; it is certainly not for him to imply that he could have been braver or wiser than they. But it was sentiments like theirs, weaknesses like theirs, and fears like theirs which helped bring Germans for a decade to be the slaves of an inhuman tyranny, which has wrecked Europe, and in its day attacked the very name of culture.

It is not Goudsmit, Morrison wrote, who should feel an unutterable pain when the word Auschwitz is mentioned,

> but many a famous German physicist in Göttingen today, many a man of insight and of responsibility, who could live for a decade in the Third Reich, and never once risk his position of comfort and authority in real opposition to the men who could build that infamous place of death.

When Meitner saw Laue's exchange with Morrison, she told Otto Hahn that he 'is not helping Germany but risks achieving the opposite'. For Morrison's last response came closest to the real point. What those who had faced the Nazis needed to hear in the late 1940s was not an explanation of how the German scientists had calculated the 'advantages' of faked compliance, nor how they had quietly got on with innocuous research under the shadow of an oppressive regime. For it was now apparent that the Nazis were not merely coercive tyrants, but perpetrators of unthinkably depraved criminality. Where then, in Laue's noble words, was there any sense of the scientists' horror at having worked within and, for whatever reasons, on behalf of a regime that gassed families and tossed their skeletal corpses on to a pile?

Lise Meitner realized this, and not just because she knew that the same fate was so nearly hers. Her response to the revelations as the Allied troops reached Dachau and Buchenwald was the true and proper one that no other physicist seemed able to adduce in their cautiously worded regrets. She simply sat by the radio and wept. 'Someone', she wrote to Hahn, interned at Farm Hall in June 1945, 'should force a man like Heisenberg and many million others to look at those camps and at the martyred people.'

In this letter Meitner felt compelled to say to Hahn things that would otherwise have obstructed the friendship that she wanted to resume:

> It was clear to me that even people like you and Laue had not grasped the real situation . . . This is, of course, Germany's misfortune, the fact that all of you had lost your standard of justice and fairness . . . You have all worked for Nazi Germany as well and have never even

tried to put up a passive resistance either. Certainly, to buy off your consciences you have helped a person in distress here and there, but you have allowed millions of innocent people to be slaughtered without making the least protest.

If we might feel inclined to respond to Morrision's remarks, despite his frank humility, that 'that's easy for him to say', the same will not do with Meitner. Not only had she experienced life under the Nazis, but she had, in her view, been complicit in it. That it should be Meitner who first acknowledges this, and not Heisenberg or Debye or poor, shattered Planck, not even Laue – this is in the end the worst of it all. 'Today I know that it was not only stupid but very unfair of me not to have gone away immediately', she told Hahn. She was tormented by that thought. But she had no illusions about her colleagues. 'You did not have any sleepless nights', she told her old colleague, asking him to read her letter only 'with confidence in my unshakeable friendship'. 'You did not want to see it; it was too inconvenient.'

Hahn did eventually see, perhaps, but it took many years. In 1958 he wrote to Meitner on her eightieth birthday, almost echoing her earlier words:

We all knew that injustice was taking place, but we didn't want to see it, we deceived ourselves . . . Come the year 1933 I followed a flag that we should have torn down immediately. I did not do so, and now I must bear responsibility for it.

He thanked Meitner 'for trying to make us understand, for guiding us with remarkable tact'. Of the fine words that were later bestowed upon scientists who worked in Nazi Germany, few speak as unalloyed a truth as those on Meitner's tombstone in Hampshire, southern England. They pronounce her 'a physicist who never lost her humanity'.

12 'We are what we pretend to be'

Kurt Vonnegut's 1961 novel *Mother Night* is the story of Howard Campbell, an expatriate American playwright who finds himself in Hitler's Germany in the 1930s. There Campbell is persuaded to make English-language radio broadcasts of racist Nazi propaganda. But unknown to the Nazis, he has been enlisted by the US War Department to lace his broadcasts with intelligence messages coded in coughs and pauses. This role is never made public, and after the war Campbell is brought to trial for his crimes. Campbell's Nazi father-in-law admits that he suspected Campbell of spying but didn't expose him because, on balance, he was more useful being allowed to continue his work anyway. He says that Campbell's broadcasts, not Hitler or Goebbels, were the inspiration for his Nazi ideals. 'You alone kept me from concluding that Germany had gone insane', he tells Campbell. The moral, Vonnegut adduced, is that 'we are what we pretend to be, so we must be careful about what we pretend to be'.*

Carl von Weizsäcker would presumably have defended a person like Campbell, for after the war he asserted that what mattered was one's intention, not one's actions. On this reasoning, the apparent support that he, Heisenberg and their colleagues had given to the National Socialists was nullified by the fact that they always disliked the government. But Weizsäcker had perhaps more reason than many to want to believe that. During the post-war denazification process

---

* *Mother Night* speaks very directly, in the way that fiction can, to the realities of living in Nazi Germany. One wonders how many scientists Campbell is speaking for when he says 'It wasn't that Helga and I were crazy about Nazis. I can't say, on the other hand, that we hated them. They were a big enthusiastic part of our audience, important people in the society in which we lived . . . Only in retrospect can I think of them as trailing slime behind.'

his father Ernst was charged for crimes against humanity as Secretary of State during the war. Weizsäcker Snr pleaded that he had stayed in his post only to aid the underground resistance to Hitler – even though by doing so he was found complicit in the deportation of Jews to the concentration camps.

Vonnegut's novel interrogates the complexities of collaboration, accommodation and resistance more searchingly than the Germans themselves were generally able to. Faced with the simplistic hero/villain narrative of Campbell's prosecutors, we are moved by the injustice of his predicament. But we cannot exonerate him because, as he himself comes to realize, the story that he told himself about his motivations during the war prevented him from ever truly questioning what the consequences of his actions were. Vonnegut calls Campbell 'a man who served evil too openly and good too secretly, the crime of his times'. He implies that we cannot invent a private self whose intentions contradict our actual behaviour, since we exist in a world of causes and effects.

Hans Bernd Gisevius, who took part in the 1944 plot against Hitler, recognized this dilemma:

> Under totalitarianism it is only possible to obstruct and oppose if one is in some manner 'on the inside'. But how far can a man participate in a hated system without selling his soul? The more the Opposition came to recognize that the Nazi rulers could be defeated only by their own methods, the harder it was for them to solve the problem of conscience. It became more difficult for them to avoid objective as well as subjective guilt. Undoubtedly many paid too dear a price for the sake of having one or both feet 'inside', and many others were unjustly accused of opportunism.

Gisevius' rather too neat categorization of modes of 'opposition' does not, however, quite tackle the difficult matter of where, if anywhere, one places the boundaries between opposition 'from inside', damage limitation with no real attempt to change the system, and merely keeping one's head down and one's hands as clean as possible.

These distinctions become all the more blurred by the insidious temptation to rearrange memory and history for the sake of self-preservation. *This* was our motivation, we insist in retrospect, and it is what we come to believe, because the illusion that we knew what we were doing is essential if we are to maintain a coherent picture of our own conscience and moral autonomy. As Nietzsche put it:

> 'I have done that' says my memory. 'I could not have done that', says my pride, and remains inexorable. Finally, my memory yields.

The Debye affair

Peter Debye had a catchphrase that his students loved to repeat: 'But you see, it's all terribly simple.' That captures in a nutshell his much-lauded ability to penetrate to the core of a scientific problem and present it in straightforward, intuitive terms. Arnold Sommerfeld averred that this was Debye's motto not just in science, but also in life. It was all so terribly simple.

And so it seemed. From the end of the war until his death in 1966, there was nothing controversial about Debye. He remained a professor at Cornell, polished his achievements but added little to them, collected awards and accolades, and maintained amiable relationships with his fellow scientists. In 1950 the DPG awarded him the Max Planck medal, which went in the previous and subsequent years to Lise Meitner and James Franck – a gesture of solidarity and reunification from the German physics community. Well after formal retirement and into his ninth decade Debye continued to attend conferences, astounding his colleagues with his stamina and eager engagement with the scientific discourse. His interjections were always insightful, always listened to and respected.* 'To the end', wrote American chemist John Warren

---

* A rare dissenter among this veneration was Paul Epstein, who was a student alongside Debye in Sommerfeld's group before moving to the California Institute of Technology in 1921. He called Debye 'not a lovable character, but very self-centered, and a great politician'. He was, said Epstein, 'not a man of the very highest professional integrity', but had a 'talent to impress people' and convinced Sommerfeld of his supposedly great abilities – 'he had him in his pocket actually'. But there is clearly some personal antipathy in Epstein's assessment – 'I saw through him, and he didn't like me for that reason' – as well as the snobbery of the aspiring middle-class intellectual:

> Debye was also a profoundly uninteresting person. That is, he had no culture; he was from a pretty low social stratum and had no general education. That is, he couldn't talk about literary problems or art problems or philosophy

Williams in 1975, 'his generosity, friendliness, and concern for others were commensurate with his mental prowess.' He gardened and went fishing, he was regarded as a family man, a devoted husband and grandfather. 'In his own eyes', says Martijn Eickhoff, 'he had got through the Third Reich . . . without a blemish. Neither Debye nor the vast majority of his contemporaries raised the question of whether the general scientific interest [although more pertinent here is surely the social and ethical interest] and his own personal scientific interest had always coincided.'

If Debye's obituaries touched at all on his position in Nazi Germany and the reasons for his departure, they followed a standard narrative: Debye had fought to minimize state interference in physics, had defended vulnerable colleagues when he could, and had left when the Nazis gave him no other option. It was sometimes suggested that he was pushed out of the KWIP to clear the way for a military takeover, not merely that he took a period of leave after the dispute about his nationality. In 1963 the *Ithaca Chronicle* insisted that he 'refused to be browbeaten by the Nazis', while an obituary by his American student Irving Bengelsdorf claimed that by leaving Germany voluntarily Debye showed 'great personal courage'.

It was agreed that Peter Debye did not care about politics but only about science. And this was presented as a virtue, or at least as a neutral position. If it made him somewhat politically naïve, there was no shame in that. Time and again his colleagues and advocates were content to leave unexamined bland comments about his 'striking lack of political interest' – as though this were no different from a lack of interest in opera, say. 'I never found in Debye any interest in philosophical questions', wrote his former associate Erich Hückel in 1972. 'Debye's way of life seemed to me rather straightforward and uncomplicated.'

It is surprising that it look so long for harder questions to be asked – and unfortunate that this was first done in so crude a manner

---

problems, and his language was somewhat pedestrian and simple . . . Debye spoke with a street accent and outlook.

One senses resentment here that such a low-born student rose so high. Debye suffered from this sort of snootiness even in his school days, when it became known that his parents were exempted from fees they could not afford.

by Sybe Rispens in 2006. As we have seen, Rispens' selective marshalling of facts to present Debye as an anti-Semite with possible Nazi sympathies and, after leaving Germany, with a determination to resume his position in Berlin at the earliest opportunity, do not survive close scrutiny.

Rispens welcomed the decisions of the universities of Utrecht and Maastricht to withdraw the use of Debye's name in the wake of his allegations. Others were outraged. The decision 'is not based on sound historical observations', said the managing director of Utrecht's Debye Institute, Gijs van Ginkel. 'I consider this decision to be faulty on the basis of our present knowledge, and I am also of the opinion that it damages unnecessarily the reputation of Professor Debye and his family, the interests of the Debye Institute, and those of the scientific community as a whole.' When van Ginkel prepared a book that attempted to clear Debye's name, the University of Utrecht halted its publication, reprimanded van Ginkel, and forbade him from talking to the press.

Many now accept that Rispens' book was misleading. Even the Dutch Nobel laureate physicist Martinus Veltman, who had contributed an appreciative foreword, realized that he had too hastily endorsed a work of questionable scholarship, and asked for his introduction to be removed from later editions. In May 2006 he wrote to the (then former) Debye Institute in Utrecht to say that

> If I had realized the consequences I would certainly have dissociated myself from the matter . . . it is now clear to me that the allegations of Rispens are unfounded and should be assigned to the 'realm of fables' . . . The question remains as to who had been damaged most by this affair. The answer is clear: the universities of Utrecht and Maastricht . . . The decision of Utrecht and Maastricht is a slap in their own face. It seems to me that the universities should admit their error, revoke their decision and further forget the matter.

Debye's family was inevitably upset by the allegations and their consequences. 'We believe you have done Peter J. W. Debye an injustice; have marred the Debye family name; and are on the verge of doing your well-known institution an enormous disservice', his son Peter and his grandchildren wrote to the University of Utrecht. In his defence they outlined his opposition to the Nazis, emphasizing the assistance he gave to Meitner and saying that he left Germany 'when it was clear that further resistance would be ineffectual'. (They do not point out that this 'resistance' was to demands that he change his nationality, not to the general policies of the regime.) 'When Debye could no longer keep politics out of his realm, he left', the Debye family contested. 'He gave up fighting the unjust from the inside. He went to the outside and helped defeat the regime he detested.'

It would be unreasonable to expect anything else from Debye's family. But it is precisely because Rispens' simplistic account of events encourages such a simplistic response, such a polarization of attitudes, that it is deplorable. This tendency even infected the Netherlands government's investigation into the allegations, conducted by Martijn Eickhoff for the Netherlands Institute for War Documentation (NIOD). Eickhoff's 2008 report, commissioned to provide an objective assessment, is often nakedly partisan and steeped in resentment and insinuation, which is all the more unfortunate because the quality of the archival research is unimpeachable. Eickhoff shapes this substantial body of valuable material into a work of pop psychology more concerned to construct a spurious motivation for its subject than to set out the facts in all their ambiguity and inconclusiveness. One wonders if Eickhoff feared that a refusal to deliver a definitive judgement would be seen as failure.

As two Dutch professors complained at the time, this eagerness to condemn or exonerate still typifies the country's position on the war years: 'On World War II we Dutch know just "good" or "wrong" – nothing in between.' Even Dieter Hoffmann and Mark Walker, the two historians who have perhaps done the most to explain the subtleties of the German scientists' responses to National Socialism, could be misinterpreted as they sought to redress the imbalanced picture painted by Rispens. They rightly pointed out that Debye's actions were entirely representative of those of many of his 'apolitical' colleagues – and yet by characterizing Debye as 'an ordinary man in extraordinary circumstances' they made his actions sound like an uncomplicated and irreproachable response to the extremes of the era.

Because of its long and hitherto proud association with Debye, the chemistry department of Cornell University launched its own inquiry into the affair with the assistance of Walker and Hoffmann. 'Based on the information to date', the department's press release concluded,

> we have not found evidence supporting the accusations that Debye was a Nazi sympathizer or collaborator or that he held anti-Semitic views . . . On the other hand, the charge that he might have been willing to accommodate the views of the Nazi regime presents a more difficult and nuanced case . . . One could also ask why he never provided an explanation or rationalization for his actions at the time . . . Clearly, we would like to have a written record by Debye detailing the rationale for his actions prior to leaving Germany. However, to suggest that the lack of such evidence is in itself incriminating is, in our view, not a defensible position.

This was not a unanimous view. The Cornell chemist and Nobel laureate Roald Hoffmann, who lost most of his Ukrainian Jewish family in the Holocaust and as a young boy narrowly escaped the same fate, was less ready to give Debye the benefit of the doubt. 'Debye took on positions of administration and leadership in German science, aware that such positions would involve collaboration with the Nazi regime', he said.

> The oppressive, undemocratic, and obsessively anti-Semitic nature of that regime was clear. Debye chose to stay and, through his assumption of prominent state positions within a scientific system that was part of the state, supported the substance and the image of the Nazi regime . . . My opinion is that Cornell should remove Debye's name from a lectureship and from a chaired professorship named after him. Debye's scientific achievements remain.

As for the bronze bust of Debye in the department's entrance hall, Hoffmann said, 'I would propose that it be moved where it belongs, into the faculty lounge.'

In a letter Debye wrote to Sommerfeld on the eve of his departure from Germany on 30 December 1939 we can find the essence of why he has been both attacked and defended. His philosophy, he explained to his former mentor, was

Not to despair and always be ready to grab the good which whisks by, without granting the bad any more room than is absolutely necessary. That is a principle of which I have already made much use.

What could be wrong with this intention to remain optimistic, looking for ways to contribute something of value and to avoid harmful actions as far as that is possible? What more could one ask? Yet one can offer another reading of Debye's words: don't attempt to change or challenge anything, but take opportunistic advantage of what comes your way while evading responsibility for the harm you do.

Which is the correct interpretation? Neither will in fact suffice, for the simple reason that Debye gives no sign of having pondered the distinction himself. His is simply a statement of shallow optimism, which will work fine – and is even praiseworthy – unless circumstances render it untenable. In Nazi Germany Debye was out of his moral depth.

It makes no sense to seek some pseudo-legalistic judgement of Debye's guilt or culpability. Eickhoff concluded that Rispens' picture of a famous scientist with 'dirty hands' was unfair, but that Debye was nonetheless guilty of 'opportunistic behaviour'. He claimed that Debye cultivated a 'principle of ambiguity' which enabled him to act selfishly in every circumstance while avoiding blame – from any direction – for the consequences. That reckoning exemplifies all that is wrong with many attempts to adjudicate on the Debye affair, for it ascribes to Debye a considered, consistent and calculated attitude that underpins his decisions. This 'principle of ambiguity' is nothing more than an elaborate way of saying that we're not sure why Debye did what he did, while making the error of assuming that Debye himself always had a moral compass to consult. In short, it refuses to accept that he – and by extension, Planck, Heisenberg, and their colleagues – was a fallible, improvising, and often unreflective human being who could not relinquish a hope that things will somehow turn out all right in the end. Klaus Hentschel laments the tendency of his fellow historians of science to disregard this aspect of human nature: few, he says, 'have the courage to relinquish the fictitiously tidy integrity of their characters'. They consider that to accept contradictory or ambiguous impulses is to capitulate, to fail in one's duty to provide a coherent account of why historical figures did what they did. Yet how often do we even know why *we* do what we do?

There are some who seem to believe that the truth of the Debye affair will emerge from yet closer examination of the archives: a diary note proclaiming sympathy with National Socialist anti-Semitism, say, or evidence that Debye was working with Allied intelligence to secure Hitler's downfall. But either possibility would be so out of keeping with every aspect of how Debye lived his public life that we would then have to regard the latter as an utter sham from cradle to grave. Debye was, in short, not that kind of person. The personality that his statement to Sommerfeld reveals is neither that of a craven opportunist nor of a brave and noble individual. It is of a man who assiduously avoided hard moral choices, and did so not by bending with the wind, but rather by cleaving to a traditional notion of duty – to science and to a system of honour – that made such choices seem unnecessary, even unwholesome. If we wish to condemn Debye for anything, it is not for his passive support of the Nazis, nor for a tacit, retrospective sanitization of his wartime actions, nor for ingratiating opportunism. All such accusations are equivocal at best. But Debye seemed reluctant to accept that a scientist has any obligations except to science. It is precisely because this has laudable as well as dangerous aspects that we find it so hard to agree on how to judge him. To deny any shades of moral greyness, however, would be to condone the picture painted in the apologia of Heisenberg, Weizsäcker and the DPG, in which the German scientists were either Nazi dupes or blameless professionals.

In personal matters Debye was a private man. We cannot be certain that he did not, on occasion, wonder if he had done the right thing in Germany. Maybe his apparent expediency and lack of concern for political matters, remarked by most of the people who knew him well, masked an inner world where he wrestled with his ethical dilemmas. Since even his family offer no evidence of that, however, it seems unlikely. In any event, it would even then scarcely exonerate him. A person who has experienced what Debye had experienced, who held positions of considerable authority in Nazi Germany and who had to make difficult choices and compromises as a result, and who has come to be regarded as something of a role model, is surely failing in their social duty if they behave subsequently as though all is well and there are no questions to be asked. Even if it were no more than a public persona, this refusal openly to interrogate the dark truths of that era is itself an act of moral irresponsibility. We can argue about the rights and wrongs of

Debye's actions in Nazi Germany; his own silence is the unequivocal failing. 'After the war', says Roald Hoffmann, 'Debye made no apology for his actions. Richard von Weizsäcker, the wise former German president, said in 1985, "*Versöhnung ohne Erinnerung gar nicht geben kann.*" There can be no reconciliation without remembering . . . I think Debye's post-war silence shows that he would have liked us to forget.'

What happened

It's not clear that a similar assessment deals adequately with either Planck or Heisenberg. Planck was always acutely conscious of doing the right thing – his difficulty was in resolving conflicting notions of what was 'right'. One can't help but feel sympathy for this man, inculcated with a deep sense of duty towards the German state and culture, when suddenly faced with a government of such criminal depravity. Planck's failure to recognize the gravity of the situation in 1933 cannot be attributed to stupidity or indifference, nor does his relatively feeble response to the encroachment of Nazi ideology seem (despite the accusations of Lotte Warburg) to have been due to cowardice. Rather, he was paralysed by a predicament for which his conservative education had never prepared him. He is, as Heilbron says, a genuinely tragic figure.

Heisenberg shared Planck's patriotic commitment to Germany and German science, with which he identified personally to an unhealthy degree. And like Debye he made his science a refuge from moral dilemmas, a higher plane that one could inhabit nobly, untroubled by the 'money-business' of politics. After the war he presented himself as a covert opponent to the Nazis, saying for example to Goudsmit that 'I knew . . . if we Germans did not succeed in undermining this system from the inside and finally to remove it, then an enormous catastrophe would break loose which would cost the lives of millions of innocent people in Germany and other countries.' Not only is it hard to reconcile such comments with his wartime remarks (for example, that one must merely wait for the Nazis' extremism to subside), but it is also difficult to understand what Heisenberg felt he was doing during that time to 'undermine the system', rather than to survive (and in some ways to flourish) within it.

What seems most to have compromised Heisenberg was a craving for approval – even that of a corrupt regime whose methods and principles he disdained – that seems concomitant with his inability to outgrow an attachment to youthful idealism. This aspect of his character surfaced in later life as an inclination towards philosophical mysticism, with which even his interpretation of quantum theory was not untouched. At the same time as insisting that his inaction and accommodation during the Nazi era was in fact the only form of 'active opposition' that could have had any effect, he sought an overblown metaphysical justification for his acquiescent conduct. The grand 'movements of thought' at such times, he said, were beyond the power of individuals to affect, and we must resign ourselves to what fate brings:

> For us there remains nothing but to turn to the simple things: we should conscientiously fulfil the duties and tasks that life presents to us without asking much about the why or the wherefore. We should transfer to the next generation that which still seems beautiful to us, build up that which is destroyed, and have faith in other people above the noise and passions. And then we should wait for what happens.

One wonders, as Lise Meitner did: did Heisenberg ever really see 'what happened'? In the 1930s the physicists already knew they were living in a thuggish, anti-intellectual state. But historians now widely agree that, towards the end of the war, any educated, well-connected person in Germany – and Heisenberg and Planck surely fit that description – will have had a good notion of the depths of its corruption: of the systematic genocide of the Jews that began in mid-1941. According to Mark Walker, 'Heisenberg knew he was working for a ruthless, racist, and murderous state.' He never condoned that, but his suggestions that it was the lesser evil and that its extremism would wane in time seem all too clearly now self-deceptions that he refused subsequently to acknowledge or even examine. Goudsmit complained to him, with some cause, in 1948 that 'not one of the German colleagues has yet denounced Nazism and pointed out how its evil features are similar to the evil side of Communism'. But that was not quite the problem. No one doubted that most of the physicists had always disliked the National Socialists, and most of them were eager to say so; denouncing the Nazis was rather easy in 1948. Yet they seemed to feel that, merely by doing so, they disassociated themselves from Hitler's regime and that this was the end of the matter.

While the formula that makes Debye 'an ordinary man in extraordinary cirumstances' risks generalizing his particular weaknesses, it rings true in the sense that there was nothing especially egregious in those failings. Debye's occasional self-interest and limited moral engagement, Heisenberg's insecurity and egotism, Planck's prevarication and misconceived notion of duty – none are profound character flaws, and all would have been minor blemishes on a fundamentally decent nature in happier circumstances. It is the grave misfortune of these men that the enormity of the conditions in the Third Reich amplified these eminently forgivable traits, transforming them ruthlessly into what some have deemed to be irredeemable faults. That is no reason to excuse actions that have profound consequences, but neither should it allow us to define the person entirely by the actions. For this is surely the sobering and indeed terrifying nature of tyrannies: that they expose us mercilessly, finding our weaknesses and bloating them out of proportion. That is why the appropriate measure of our conduct is perhaps not so much what we did as how we deal with it subsequently.

Are scientists special?

Is there any reason to expect from Planck, Heisenberg and Debye something more than their compromised, halting and ambivalent moral stance, purely because they were scientists? Did their positions as leading members of the German physics community create obligations and expectations any more demanding than those one might impose on the general population? There is a widespread view that scientists are no more morally accountable than the rest of us. That is mostly a valid claim, although situations may surely arise – the development of nuclear physics is one such – in which the superior knowledge that scientists possess confers a special duty to consider the wider social and political implications of their research: they alone can evaluate how it might be used and abused. But the broader question is how morally aware and responsible the professional institutions and attitudes of science are.

We have seen how it was a common belief among German scientists between the wars that the proper and noble conduct of their profession entailed an 'apolitical' withdrawal from the messy, compromised power struggles of civic society into the realm of logic, abstraction and 'truth'. Because he engaged with worldly affairs, Einstein was condemned sometimes even by those who revered his work for 'making science political'. This conviction can still be detected in researchers today. Scientists pride themselves on offering facts, not opinions, and some insist on drawing a distinction between the purity of scientific discovery and the dirty realities of its application. To the public, this disengagement from the realities of commerce, societal considerations and politics is apt to make the scientist appear like an amoral Dr Strangelove.

The naïvety of such simplistic postures was exposed in Nazi Germany. On the one hand, an 'apolitical' stance left the scientists vulnerable to political manipulation; indeed, it became in itself a politically implicated position, since being apolitical prevented one from directly criticizing the government. At the same time it was a facade, for the scientists used the bait of nuclear power to extract funds from a somewhat sceptical regime – and if they did not get more than they did, it was because they lacked conviction in the real potential of their own research, and not because the money was not there for the taking. Few scientists would today deny that it is something of a game to obtain support from governments and companies increasingly interested in the short-term financial return from their investment in research. But they are more reluctant to accept that this makes science itself political and morally accountable. And this is not just because scientific discoveries have social consequences, but because the scientist becomes a player in the political landscape. Indeed, it is the very humanitarian motivation of a great deal of science, from drug research to energy technology, that gives it a moral and political orientation. If scientists wish to do good in the world – and most of them certainly do wish this – then they must see that this makes the very act of doing science a political one.

Evasions, delusions, diversions: these were how most scientists accommodated themselves, usually unwillingly and often unwittingly, to National Socialist Germany. As Alan Beyerchen says, 'the truth was not that the scientists were political cowards, but that they did not know how to be political heroes'. Their vision was too narrow, their standards too conservative. It was not so much that these men blindly followed a redundant notion of duty, but that they seem actively to have constructed an idea of 'duty to science' as a way of denying broader responsibilities. As Debye put it in a letter to his compatriot Pieter Zeeman in Amsterdam in 1937, 'It is always my custom to ask myself in what way I can be most useful for physics. That is the first consideration for me and other more personal considerations play a more secondary role.'

Ironically, it was Debye himself who pointed out that the German scientists sought refuge in their work. After his meeting with Warren Weaver in New York in February 1940, Weaver reported that

> Debye comments that Hitler accomplished, by going to war, a complete identification of himself with Germany. Under peace, an intelligent citizen can perhaps distinguish between Hitler as an individual, his policies and principles, and the Fatherland. But in time of war this distinction goes by the boards. Debye says, for example, that he knows any number of fine, intelligent Germans who are working to the very limit of their capacity and energy (and it is a very high limit) on the specific jobs which have been assigned to them. Such persons find a sort of emotional relief in having a job in which they can work almost to exhaustion. They do not stop to question, or feel that it is possible to question, any matter of broad policy or general direction. What they do now, in time of war, they do not do for Hitler but for Germany. It is neither possible nor proper to worry about general policies; one only has to do his own individual job to the very limit of his ability.

This intention to work only 'for science', regardless of political or moral issues, troubled some commentators in the post-war nuclear age. The Swiss playwright Friedrich Dürrenmatt examined the moral dilemmas of the nuclear physicists in his satirical 1962 play *The Physicists*, in which three physicists incarcerated in a lunatic asylum offer different views on how to reconcile their work with their responsibilities. One avows allegiance to his nation; another insists that 'we have far-reaching pioneering work to do and that's all that should concern us'. All three are guilty of overestimating the influence on politicians of their own opinions about how the 'new and inconceivable forces' they have unleashed should be used – as with many scientists of the early twentieth century, including Heisenberg and the hapless Bohr, their grand schemes of a new world order guided by scientific sages will barely even reach the contemptuous notice of their political leaders.

What could they have done?

How easy and how tempting it is, though, to condemn the German scientists by an accumulation of compromising particulars. One might say that Planck could have stood up to a knee-slapping Hitler, that he could have supported Einstein rather than requesting his resignation from the Prussian Academy of Sciences. Heisenberg might at the very least have desisted from advising his former colleagues in occupied Denmark that they must reconcile themselves to a German victory; he might have thought twice about giving propagandizing talks that polished his own profile in Nazi circles. Debye could have resigned his leadership of the German Physical Society rather than sign off that fateful letter with 'Heil Hitler!'

But without the benefit of hindsight from a safe and comfortable viewing seat, what might we reasonably have expected the physicists to do differently? Hans Bernd Gisevius avers how hard it was to take an individual stand against the regime:

> Let us not forget that totalitarianism and opposition are two mutually exclusive political ideas. In a democracy it is possible to practise opposition, but dictatorship permits no antagonists; it does not even put up with the lukewarm and the sceptical. Whoever is not for it is against it. Oppositionists must keep silent, or they must decide on underground activity.
>
> Underground resistance and opposition are again two different matters. Opposition is struggle against an existing regime; it is an attempt to bring about a shift in course or a change in personnel without directly overthrowing a system. Opposition, therefore, recommends a more prudent policy, offers reasoned advice, tries to reform by appeal to the common sense of the rulers and attempts to win the favour of the voters; but the oppositionist under a totalitarian system must not try to reform at all. His good advice would only help the tyranny; any intelligent recommendation would support the reign of terror.

Gisevius' remarks have been used to justify the relative complacency of the German scientists' response to Hitler. The only alternative, they seem to say, was to do as Gisevius did and plot the violent destruction of the Nazi leaders, risking their own certain and immediate death. But Gisevius makes too extreme a case. A few scientists, such as Laue and Fritz Strassmann, were openly 'lukewarm and sceptical' about the regime – indeed, rather more than that – and yet they were, in some degree or another, 'put up with'. Such dissent was not necessarily suicide, even professionally, although it could undoubtedly cause one trouble.

During the war itself, it was a somewhat different matter. Asking himself in his memoirs why he had not joined the Dutch resistance or given more aid to the Jews, the Dutch physicist Hendrik Casimir showed commendable and even rather moving honesty:

> I felt I had been a coward and an opportunist . . . I had on occasion, for short periods, given shelter to people who had to hide and I had once or twice just escaped being arrested. It was not enough . . . I had been afraid of having to face human cruelty, of having to face the risks of being questioned and tortured . . . I was not cut out for the "illegal" underground work.

He had always tried to avoid disagreements and conflicts, he said, but 'during the war that was the wrong attitude'. So, Casimir concluded, 'I think my behavior can be explained and, perhaps, partly excused, but even today that does not entirely remove my feeling of guilt.' Whatever we feel about Casimir's confession – and there is surely some moral bravery here to compensate for his self-avowed lack of physical bravery in wartime – such soul-searching is notably absent from almost all of the physicists who, not even the victims in an occupied country, actually worked in Nazi Germany and in some cases profited from it.

Far more problematic than the rarity of Laue-like opposition are two more general features of the scientists' attitudes. The first is the almost total lack of a moral position. On several occasions, Planck, Heisenberg and Debye all showed courage in refusing to comply with political demands. But there is no evidence that in these cases their behaviour was informed by a broad moral perspective. They helped Jewish colleagues because they were colleagues, and not because they deemed it perverted to oppress and expel Jews. And they deplored this oppression not because it was inhumane but because it would damage German science. In resisting, they tended to defend not a moral principle but their own autonomy and traditions. Planck was determined to honour Haber not because this would symbolize resistance to anti-Jewish prejudice but because a failure to do so would violate his code of professional duty.

This does not mean that the scientists were blind to the inhumanity of anti-Semitism, and it would be unfair to imagine they were indifferent to it, let alone that they condoned it. But it testifies to a limited concept of where virtue lay. In this, the scientists' stance was no different from that of much of the German population who did not actually applaud the anti-Semitic measures. Being scientists did not make these men any less sensitive to the plight of Jews; neither did it give them any greater ethical sensibility. Rather, it enabled them to convince themselves that adherence to professional standards, as far as was possible, was a form of 'opposition'. But Beyerchen is right to conclude that in fact this 'was not opposition at all . . . in an environment like that created by the Third Reich, political opposition is the only opposition worthy of the name'. By assuaging consciences to no real effect, 'professional opposition' was, as Gisevius says, arguably worse than useless.

Second, and perhaps most troubling of all, there was an almost universal inability among the scientists to acknowledge or even recognize their failures in retrospect. It is one thing to display poor judgement, lack of resolve, or self-interest in a crisis; indeed, it is normal. It is another to express no remorse later – more, to reconfigure the historical narrative so that remorse is not even demanded. Nothing can excuse Carl von Weizsäcker's suggestion at Farm Hall – which he never recanted – that the German nuclear scientists, threatened by a ruthless dictatorship, 'obeyed the voice of conscience' while the Allied scientists, with nothing to fear, created a weapon of immense destructive power. What is most alarming is not that the scientists sought to justify their actions, which is after all a universal human weakness, but that in some ways they did not even imagine any such justification was necessary. I believe that Peter Debye would have been surprised and astonished at the accusations made against him in 2006. He would have been right to consider them ill-posed and largely unjust, but wrong to be dismayed that such questions might ever be asked.

In 2011 the Debye Prize was awarded in Maastricht for the first time since 2004. It was given to the director of the reinstated Debye Institute in Utrecht, and was presented in the town hall where the bronze bust of Debye still presides. This seems a defensible outcome, since the decision to remove Debye's name from these institutions was an ill-considered and reactionary political gesture that did no one any credit. But this does not mean that the question of Peter Debye has been resolved in his favour. Nor should it be.

Indeed, one minor consequence of the Debye affair is that it should prompt some reconsideration of the practice of naming institutions after 'great scientists'.* The motivation is questionable at best. In response to the Debye affair, historian Leen Dorsman of Utrecht University lamented the 'American habit' of naming institutes for individuals: 'The motive is not to honor great men, it is a sales argument. The name on the facade of the institute shouts: Look at us, look how important we are, we are affiliated with a genuine Nobel laureate. It is now clear from this sequence of events that this raises problems. The stakes are so high that panic reactions (on both sides) are the logical consequence.' The practice is widespread in academia, but science seems peculiarly keen to canonize its 'greats' in this way. The intent is undoubtedly not always as ignoble as Dorsman implies, but evidently scientific eminence alone is not the determining factor: there is no longer a Philipp Lenard Institute at Heidelberg. If that is the case, then such accolades impose an unrealistic expectation of probity on the part of those so honoured. Scientists rightly insist on a distinction between the quality of one's science and the quality of one's character. If so, why create a situation where the two must necessarily be conflated?

The sequence of removing Debye's name from an institute and

---

* The Alexander von Humboldt Foundation awards a Werner Heisenberg Medal for promoting international collaboration, the irony of which is hard to deny in the light of Heisenberg's wartime propaganda lectures in occupied countries, while the German Research Foundation (DFG) awards Heisenberg professorships. The DPG's Max Planck Medal for outstanding work in theoretical physics presents a much more conciliatory prospect: between the end of the war and 1970, many of the recipients were of Jewish descent, including Max Born (1948), Lise Meitner (1949), Gustav Hertz and James Franck (1951), Rudolf Peierls (1963) and Samuel Goudsmit (1964). Einstein was the first recipient in 1929. The medal was awarded to Heisenberg in 1933, Debye in 1950, and Weizsäcker in 1957. There is, to my knowledge, no Max von Laue Award.

then reinstating it implies a process of disgrace and rehabilitiation – of verdicts of guilt and then innocence. That is precisely what, in the cases of Debye, Heisenberg, Planck, and many others in Nazi Germany, we must seek to avoid. For by simplistically condemning or absolving them, we abrogate responsibility for the dilemmas that science and scientists face, always and everywhere.

Epilogue:
'We did not speak the same language'

A Nobel laureate recently suggested to me that the Nazi atrocities were a consequence of 'religiosity', which all good scientists – as torch-bearers of the tradition of Enlightenment rationalism – should reject. While both the historical validity and the logic of this claim are as warped as that of Pope Benedict XVI's suggestion that the Nazi tyranny was a consequence of 'atheist extremism' (there is no surer or more facile way to win an argument than by placing Hitler in the opposition), nevertheless it expresses a common notion among scientists that their calling should insulate them from the excesses of ideologies of all sorts. Scientists, Otto Hahn claimed with breathtaking arrogance in 1947 as he railed against the iniquities of denazification, are 'accustomed to regarding matters perhaps a little more calmly and rationally than other professions'. Science historian Joseph Haberer concluded in 1969 that 'an idealization of science as a superior form of activity remains deeply entrenched in the contemporary scientific consciousness'. One can safely make the same statement today.

The dangerous complacency of this assumption is laid bare by the history of German science under National Socialism. It should be obvious from even a cursory consideration of the matter that the rational and impersonal viewpoint required in science here conferred absolutely no advantage in matters of morality. Indeed, the behaviour of German physicists in the 1930s shows that the situation is potentially worse still. While several German religious leaders, writers and artists, industrialists and politicians mounted strong opposition to Nazi rule at great personal cost, sometimes of their lives, there was nothing comparable to be found in German science.

This was not, on the whole, because the scientists sympathized with the regime, even if like many middle-class liberals they might initially have agreed with some of its general principles, such as nationalism, robust leadership and foreign policy, and a reduction in the influence of Jews in public life. All the same, the scientists' later insistence that colleagues who were ardent supporters of the Third Reich were aberrations, mediocrities or lunatics must be seen as an attempt to 'cleanse' the scientific profession of ideological taint. Yet as Haberer has said,

> The real issue involves how it was possible for men trained in the sciences, like Lenard and Stark, to become fanatical National Socialists. If Nobel laureates can be so infected, what protection does scientific training and practice provide against the excesses of irrational personal, economic, social or political conduct? Most scientists have tended to assume that they (more than any other professional type) follow the paths of rational, disinterested, and even humane conduct. The evidence increasingly demonstrates that scientists as a whole are no more immune to the ailments of political man than other men.

This is more even than a matter of saying that scientists are no better, morally speaking, than the rest of us – a conclusion that should surprise no one, despite the delusion of some scientists that reason and moral virtue go hand in hand. For while Stark and Lenard were indeed in a minority, many scientists found in their profession a justification for *avoiding* questions of social justice and probity: their duty was only to science. Thus, Haberer implies, while there is no reason to expect science to be any more principled than other areas of human activity, it's possible that it may be *less* so. 'Whether they support the regime or not', a group of science historians has recently written, 'most scientists, or perhaps better put, scientific communities, will do what they have to in order to be able to do science.'

All the same, science does have a tradition of liberalism. Today scientists of almost any nationality tend to be more internationalist, more tolerant, more left-leaning and progressive, than a cross-section of the population from which they are drawn. But this is probably more to do with the culture that has evolved in post-war science, and among the educated intelligentsia generally, than with scientific training per se. In similar fashion, it was the background and professional development of the German physicists, not their science, that dictated their responses to Nazi rule: their conservatism, patriotism and sense of duty. There have been and continue to be among scientists some individuals with a strong commitment to global peace, such as Joseph Rotblat and Linus Pauling, as well as prominent political dissidents such as Fang Lizhi and Andrei Sakharov – but these are generally brave, principled people who just happen to be scientists (and who owe their political voice to that fact).

We must also distinguish opposition to state interference in order to protect the scientific profession from expressions of broader social conscience. Many scientists are frequently and rightly outspoken today about infringements of the freedom of speech, and will steadfastly support oppressed colleagues working in authoritarian regimes. But defending the rights of one's peers doesn't always entail an acknowledgement of the wider moral issues. I once attended a session on human rights during an international physics conference in Paris – itself a highly commendable rarity at such an event – at which the panellists spoke eloquently and passionately on behalf of scientists imprisoned for challenging their political leaders, but fell silent when asked about the legitimacy of weapons research in the light of the clear link between arms trading and human-rights violations. To address that matter would mean to infringe on colleagues' freedom to choose the direction of their research.

Moreover, championing 'free speech' – in principle an asset to the scientific enterprise that is rightly treasured – may become a reflex formula that trumps any other moral judgement. When a lecture at London's Science Museum by the Nobel laureate biologist James Watson was cancelled in 2007 after Watson made racist remarks about intelligence in a newspaper interview (he claimed that 'people who have to deal with black employees' know the assumption of equal intelligence among races to be untrue), biologist Richard Dawkins protested at 'the hounding, by what can only be described as an illiberal and intolerant "thought police", of one of the most distinguished scientists of our time'. Not only did this fail to recognize that Watson was using his privileged platform to voice anecdotal prejudice rather than a scientific hypothesis, but it implied that his professional standing as a scientist should in itself offer some protection against censure.

Without wishing to draw too lurid a parallel, one can't help being reminded of Arnold Sommerfeld arguing that the judgement of the Nuremberg trials on Johannes Stark be mitigated by his 'scientific importance'.

Scientists often say that they cannot be expected to be proficient in making moral and ethical judgements as well as technical ones. This position was adduced by the American physicist Percy Bridgman in an article on 'Scientists and Social Responsibility' in the March 1948 issue of the *Bulletin of the Atomic Scientists*. Bridgman argued that the social consequences of research must lie outside the scientist's domain. After all, how can scientists possibly expect to foresee the ways in which their work will be applied, let alone then ensure that only beneficial uses are pursued? Either they would be regulated and constrained beyond measure, not to mention legally vulnerable, or they would be paralysed by bureaucracy. They are not, in any case, trained to be competent in areas of ethics or public policy.

In fact Bridgman's view was rather more extreme. He considered that the demands of science make it necessary for scientists to be freed from the shackles of moral or social constraints altogether, so that they have no obligation to consider what consequences their work might have:

> The challenge to the understanding of nature is a challenge to the utmost capacity in us. In accepting the challenge, man can dare to accept no handicaps. That is the reason that scientific freedom is essential and that the artificial limitations of tools or subject matter are unthinkable.

Most scientists today might be hesitant to express such a forthright view, although I have no doubt that some would defend it with passion, and others would secretly find it alluring. Certainly, many are happy to proclaim the simplistic notion that 'there are no questions that should not be asked' – forgetting what politicians and the media show us every day, which is that the mere framing of a question can be a politically freighted act.*

---

* Mark Walker has pointed out to me the echo here of the assertion by the conservative historian Ernst Nolte that 'there are no forbidden questions', made in the context of his revisionist analysis of the Holocaust during the 1980s, wherein he attempted to exonerate the Nazis by comparing their genocide with that of other regimes and nations.

Although Bridgman is right to say that scientists have no special moral competency, the statement is somewhat self-fulfilling. Scientific training rarely incorporates an ethical dimension. Even when it does, the emphasis tends to be solely on codes of professional conduct: issues such as intellectual property, citation, treatment of staff, conflicts of interest and whistle-blowing. One might also ask whether an essentially technical vocation should incur any more moral responsibility than that expected of the average citizen: car mechanics and chefs, say, are not troubled by such demands. But it seems proper that one's obligations in this regard should follow in proportion to the potential impact of one's actions. The development of nuclear weapons during the Second World War brought this issue to a head by revealing how socially and politically transformative, not to mention how destructive, a new technology can be.

In the light of developments such as genetic engineering and nanotechnology, there is far greater awareness today that new technologies raise important societal and ethical questions that should be debated within and beyond the scientific community in parallel with their technical development. It is also generally recognized that scientists themselves cannot be expected to anticipate all such problems and dilemmas, or to adjudicate them alone. Yet this has not necessarily bred a readiness in scientists to engage with these matters beyond the role of offering technical advice. A common response is to acknowledge that these are important questions but to insist that they must be left for 'others', or for 'society', to decide – that the accountability of scientists extends only to matters of technical judgement and the objective presentation of data and evidence.

The limited ethical horizons of much of the scientific community go hand in hand with a conscious disengagement from politics. The belief that science should somehow be 'above' politics has been evident at least since the inception of modern science in the seventeenth century. Yet at the same time, that historical perspective also shows how ineluctably science has been bound to politics, not least in terms of the scientific community's need for state sanction and support. The practice of science, says Haberer, 'is infused with problems which require political modes of thought and political instrumentalities'.

A reluctance to embrace this aspect of science has meant that its community has generally not distinguished itself in the political arena.

Compared with the clashes that have arisen between governments and some artistic and religious movements, Haberer claims that 'scientific leadership has tended, almost without exception, to acquiesce in any fundamental confrontation with the state, especially when opposition was likely to evoke serious sanctions'. And individual scientists have often displayed a misplaced conviction that they can manipulate state leaders for their own ends, only to find that it is they who are used and then discarded. Even scientists who do show moral courage are prone to this mistake. There can be few more poignant scenes in the history of nuclear proliferation than Niels Bohr's disastrous audience with Winston Churchill in which he hoped to convince the British prime minister of the need to engage in frank dialogue with the Soviets about atomic weapons. C. P. Snow described that meeting as 'one of the blackest comedies of the war', in which Bohr and his son Aage were sent away with a flea in their ear. 'He scolded us like schoolboys', Bohr said afterwards. 'We did not speak the same language.'

While one can't expect scientists to be braver or more morally astute than any other section of the population, science can and should as a community organize itself to maximize its ability to act collectively, ethically and – when necessary – politically. That objective would need to include more explicit recognition of the political nature of science itself, and should relinquish its reliance on unexamined myths about 'scientific martyrs' to ideology such as Galileo or (so the conventional story goes) Giordano Bruno.

When science does confront politics, it has often been apt to do so with a kind of naïve, Platonic view in which political action is conducted in some abstract sphere where questions of right or wrong hardly exist. The German psychiatrist and philosopher Karl Jaspers detects this baleful tendency in Robert Oppenheimer's pronouncements on the social roles of science, full of imagery of the statesman practising his skill of 'statecraft' upon the body politic while failing to locate any particular nexus of moral choice. The scientist, meanwhile, wanders in innocent awe among nature's marvels, detached from consequences. As Oppenheimer put it:

> We regard it as proper and just that the patronage of science by society is in large measure based on the increased power which knowledge gives. If we are anxious that the power so given and so obtained be

used with wisdom and with love of humanity, that is an anxiety we share with almost everyone. But we also know how little of the deep knowledge which has altered the face of the world, which has changed – and increasingly and ever more profoundly must change – man's views of the world, resulted from a quest for practical ends or an interest in exercising the power that knowledge gives. For most of us, in most of those moments when we were most free of corruption, it has been the beauty of the world of nature and the strange and compelling harmony of its order, that has sustained, inspirited, and led us. That also is as it should be.

While Oppenheimer's statement does speak to the honourably idealistic impulse that motivates many scientists, it is at the same time a sweetly worded diversion from the issues and a misrepresentation of the daily business of science – another myth of its apolitical character. In contrast to Oppenheimer's rose-tinted view, scientists in fact rarely miss an opportunity to point out the possible applications of their discoveries. If we now deplore Paul Harteck's and Heisenberg's efforts to win military funding or political prestige by parading the possible uses of nuclear physics, it is not because of those appeals in themselves but because they were directed to the Nazis. Oppenheimer's comments on the alleged moral neutrality of science – words strikingly similar to those voiced by Peter Debye – take on a very different complexion when read against the context of German physics in the 1930s, as he more than anyone should surely have known:

In most scientific study, questions of good and evil, or right and wrong, play at most a minor and secondary part . . . The true responsibility of a scientist, as we all know, is to the integrity and vigor of his science. And because most scientists, like all men of learning, tend in part also to be teachers, they have a responsibility for the communication of the truths they have found.

Well might we then understand Jaspers' complaint:

We hear different language from a scientist like Oppenheimer . . . talking of 'beauty', or our faculty of seeing it in remote, strange, unfamiliar places or paths that maintain existence in a great, open, windy world . . . This is the premise of man, and on those terms we can help, because we love one another. In such sentences I can see only an escape into sophisticated aestheticism, into phrases that are existentially confusing, seductive, and soporific in relation to reality.*

Seen against the wider historical backdrop, the behaviour of German physicists under the Nazis was evidently not an aberration under extreme circumstances but rather, a fairly typical example of how science and politics interact. As Mark Walker says, 'It must be possible both to respect the unique, terrible nature of National Socialism [in Germany] and compare it with other periods in history.' Historian Kristie Macrakis is surely right to claim that 'Many of the ways in which the social order influences science in turbulent times are present in dormant forms in science organizations, science policy, and the practice of scientific research in normal times, or in a democracy.' And while of course this particular episode cannot illuminate or exemplify all aspects of how scientists operate morally and politically, nevertheless such case studies are a more trustworthy gauge of how science functions within society than general assertions about the 'scientific attitude'. Robert Oppenheimer's vague ruminations about 'statecraft' tell us far less about how scientists and politicians interact than the McCarthyite realpolitik that stripped him of security clearance and authority in the 1950s.

In this respect, the lesson is not that the German physicists, as a group, failed to offer sufficient opposition to Hitler. That conclusion is hard to deny, but it is a brave person who asserts without hesitation that he or she would have done better, shown better judgement, been braver, had a clearer view of where choices would lead. Rather than simply accusing them of being morally wanting, Haberer draws a more valid and much more general judgement:

---

* Oppenheimer's famous remark that after Hiroshima 'the physicists have known sin' seems in contrast to be an admission of guilt. But it too is ambiguous, not least in the elusive tense of 'have known'. And did they, moreover, not know it until 1945? In this way, Haberer says, such comments 'elude specific meaning'. Even Oppenheimer's biblical phrasing is arguably a shield against the immediate, very practical questions that nuclear power raises.

The failure of scientists has lain in their moral obtuseness, in their incapacity to define, delineate or even to recognize the nature of the problem of responsibility. Characteristically, responsibility has been recognized only in its narrower sense. Scientists have been willing to be held responsible for the calibre of their scientific work; or when acting in administrative positions for their performance in terms of the formal responsibilities attached to their positions. Beyond this methodological and bureaucratic responsibility scientists have not, at least until very recently, ventured.

For the choices they made, I do not judge Debye, Planck or even Heisenberg as harshly as some have done. But it is very hard indeed to see how they can be exempted from the failure that Haberer here describes. In this, they were representative of most scientists of their times.

A new dialogue

If the community of science today does not wholly escape these charges either, we would nevertheless be mistaken to suppose that nothing has changed. The Manhattan Project and the nuclear arms race that followed played a big part in cultivating a recognition of wider responsibility. So too have many other episodes since then, among them environmental despoliation and climate change, thalidomide, the link between smoking and cancer, genetic engineering, Chernobyl, AIDS, embryo research and synthetic biology. It is unfair to suggest that science continues doggedly to insist on its abstract purity and detachment from morality.

Scientists' acceptance of responsibility in these and other instances has, however, sometimes emerged only under duress. The public backlash against genetically modified organisms in the 1990s, for example, forced researchers in this field to address the need for dialogue, or in the jargon of our day, 'public engagement'. It is also true that some of this 'engagement' is prompted more by a desire to avoid overly restrictive and poorly informed regulation than from a profound wish to develop principles of good conduct. But it would be churlish and cynical to suppose that this is as far as the matter goes.

The launch of the *Bulletin of the Atomic Scientists* in 1945 by some of the researchers involved in the Manhattan Project was the first indication after the war that science was ready to acknowledge its social and ethical obligations. The magazine was an explicit attempt to counteract political abuses of nuclear physics and to alert the wider world to the dangers of the new knowledge. In 1947 it introduced an iconic symbol to convey these perils: the Doomsday Clock, on which the proximity of the hands to midnight illustrates the scientists' consensus on the danger of global nuclear apocalypse. Today the *Bulletin* has broadened its focus to other potentially catastrophic dangers of science and technology, in particular climate change and new technologies in the life sciences. In 2007 the Doomsday Clock was moved from seven to five minutes from midnight in response both to the existence of many thousands of nuclear weapons in an ever-growing clique of nations and to the destruction of human habitats from climate change.

The readiness of nuclear scientists to shoulder their onerous responsibilities was also signalled by a gathering of scientists in 1957 at a meeting in the village of Pugwash in Nova Scotia, Canada, to discuss the proliferation of nuclear arms and the escalation of tensions between the Soviet Union and the West. That meeting, sponsored by the Canadian banker and philanthropist Cyrus Eaton, was triggered by the release two years earlier of a manifesto written by Bertrand Russell and Albert Einstein, calling for scientists to 'assemble in conference to appraise the perils that have arisen as a result of the development of weapons of mass destruction' and appealing for peaceful reconciliation of East and West. The manifesto was signed by, among others, Max Born, Percy Bridgman, Frédéric Joliot-Curie, Linus Pauling and Joseph Rotblat. The Pugwash meeting was the first in an ongoing series of conferences on 'science and world affairs', focusing in particular on nuclear weapons, chemical and biological warfare, and international diplomacy.

In 1995 Rotblat and the Pugwash organization were jointly awarded the Nobel Peace Prize. In his address, Rotblat condemned the 'disgraceful role played by a few scientists . . . in fuelling the arms race'. He quoted with approval the words of anatomist Solly Zuckerman, chief scientific adviser to the British government from 1964 to 1971:

When it comes to nuclear weapons . . . it is the man in the laboratory who at the start proposes that for this or that arcane reason it would be useful to improve an old or to devise a new nuclear warhead. It is he, the technician, not the commander in the field, who is at the heart of the arms race.

Rotblat called on his fellow scientists to relinquish the myth of 'apolitical science' and to face the dilemmas that their research creates:

You are doing fundamental work, pushing forward the frontiers of knowledge, but often you do it without giving much thought to the impact of your work on society. Precepts such as *'science is neutral'* or *'science has nothing to do with politics'* still prevail. They are remnants of the ivory tower mentality, although the ivory tower was finally demolished by the Hiroshima bomb.

Rotblat's words belie any notion that scientists refuse to embrace moral questions, but at the same time they illustrate that even in recent times such an acceptance of responsibility is not the norm.

Another important acknowledgement of the scientist's ethical duties occurred in 1975, when many leading biologists gathered at the Asilomar Conference Center in Monterey, California, along with members of the press and US government, to discuss the implications of new techniques in genetic engineering: the ability to excise and insert genes into DNA. Such methods are now one of the dominant influences on molecular biology, being central not only to the creation of genetically modified organisms for research, agriculture and breeding, but also to new forms of medicine (gene therapies), cloning, and genomic profiling. As one attendee, the Nobel laureate biochemist Paul Berg, has put it, 'Looking back now, this unique conference marked the beginning of an exceptional era for science and for the public discussion of science policy.' Scientists had become aware that, while genetic engineering created extraordinary opportunities in medicine, industry and fundamental research, it also had serious risks. Some felt, according to Berg, that 'unfettered pursuit of this research might engender unforeseen and damaging consequences for human health and the earth's ecosystems' – and that as a consequence there should be a voluntary moratorium on certain avenues of research.

The Asilomar conference did not recommend such a moratorium, but instead led to the imposition of strict guidelines on the new genetic technologies. This 'cautious permissiveness' seems now to have been a wise position, for the worst fears about public health hazards have not materialized despite the many millions of experiments that have used the techniques. In Berg's view, Asilomar was a success not only because it made the right decision for science but also because of its impact on the image of how science is done:

> First and foremost, we gained the public's trust, for it was the very scientists who were most involved in the work and had every incentive to be left free to pursue their dream that called attention to the risks inherent in the experiments they were doing. Aside from [the] unprecedented nature of that action, the scientists' call for a temporary halt to the experiments that most concerned them and the assumption of responsibility for assessing and dealing with those risks was widely acclaimed as laudable ethical behavior.

This is a matter of public relations, but not just that: an increasingly suspicious public (and that suspicion, the tarnishing of science's halo, began with Hiroshima) will not be easily fooled by scientists going through the motions of a societal duty to which they aren't genuinely committed. However, although Asilomar demonstrated a commendable readiness to consider consequences and accept inconvenient conclusions, Berg doubts whether the same approach will work today for some of the ethical issues raised by genetic and biomedical research, such as embryo research and stem-cell technology. It is one thing to evaluate objective health risks, even though this alone is hard enough in the face of unknown consequences and the vagaries of public risk perception. But when science confronts deeply held social and religious values, it is far from clear that a consensus can ever be reached, even by compromise. Society has to find some way of accommodating irreconcilably different views. It is neither science's duty nor its prerogative to resolve such questions. But we should hope that it continues to cultivate a community in which an awareness that they must be confronted is found not just in a few unusually thoughtful individuals.

Science and democracy

While German National Socialism cannot stand proxy for every autocracy in the modern world, the fate of science under its auspices challenges some preconceptions about the relationship of research and political democracy. Many Western scientists cleave to the idea that science can only truly flourish in a wholly free society (forgetting that it did not arise in one). This was Samuel Goudsmit's agenda in his attacks on Heisenberg, who he wanted to push into admitting – without strong justification – that the Germans failed to make the bomb because Nazi interference in German science had left it too enervated. The attitude is evident too in the common perception among scientists that the Nazi leaders rejected aspects of modern theoretical physics on the ideological grounds that they were 'Jewish'. As we have seen, the National Socialists were more pragmatic than that – they lost interest in 'Aryan physics' when it became evident that it was merely a distraction from, and perhaps a hindrance to, useful technologies.

This attitude that only democracies can and will nurture science is unduly and perhaps dangerously self-congratulatory. The work of historians of science, such as Yakov Rabkin and Elena Mirskaya, dispels that illusion: 'The history of science in totalitarian societies', they say, 'makes associations between science and freedom appear tenuous at best.' Not only have such regimes often been quite generous in their support of science, but the scientific attitude of detached objectivity can be and has been adopted by these regimes to legitimize regarding their own citizens in the same way. Mark Walker and a group of other eminent science historians agree that 'no single ideology, including liberal democracy, has historically proven more effective than another in driving science or leading to intended results'. The project commissioned by the Max Planck Society in the late 1990s to look into its (that is, the KWG's) murky past concluded that the society as a whole did not simply 'survive the swastika' but was successful in pushing its own agenda and in some ways flourished under Hitler. Indeed, it deemed that 'the Kaiser Wilhelm Society was an integral part of the National Socialist system of domination that subjugated people inside and outside Germany and culminated in genocide and war'. Challenging the idea that science and mathematics are inherently democratic, historian Herbert Mehrtens argues that 'they will adapt to political and social changes as long as there is the chance to preserve existence'. After examining closely the history of mathematics in Nazi Germany, Mehrtens concludes that 'I cannot find any reason why mathematics, and any other science, should not find a perfect partner in technocratic fascism.'

The common suggestion that a non-democratic country might ape the innovation of the democratic West but can never match its scientific creativity is an arrogant delusion. Even during the height of the Cold War, when state oppression in the Soviet Union was more extreme than it was in pre-war Germany, Soviet scientists were capable of inventive and effective scientific research. And today Chinese scientists are increasingly proving that, even in the face of the rote learning of China's traditional education system, democracies have no monopoly on creativity. This should surprise no one. Most Chinese scientists today enjoy considerably more state support, personal liberty and freedom from demagoguery than most German scientists had done under Nazi rule, yet even the latter were perfectly able to conduct vibrant and productive science, not least the work that led Hahn and Strassmann to discover nuclear fission in 1938.

Many scientists believe that dictatorships will inevitably constrain science by imposing an arbitrary ideology on what may or may not be discovered and taught. That has certainly happened: one of the most notorious ideological distortions of science was Stalin's suppression of Darwinian genetics in favour of the agriculturally disastrous Lamarckian views on heredity propagated by Trofim Lysenko between the 1920s and the 1960s, which Lysenko had couched in a politically expedient Marxist framework. But this sort of interference is rare. We have seen how 'Aryan physics' enjoyed rather little support from the Nazi government, largely because those leaders who recognized the value of science did not believe it could deliver the goods. 'No political regime has ever tried consistently and comprehensively to impose ideologically correct science on its scientists', say Walker and colleagues – in part because 'the military potential of science and scientists outweighs and overrules attempts to purify science ideologically'. Hitler was prepared to slacken the shackles of anti-Semitism for pragmatic ends during the war, and not even Stalin risked the politicization of nuclear physics. 'Stalin left his nuclear physicists alone', says historian Tony Judt. '[He] may well have been mad but he was not stupid.'

That science and democratic values are uncoupled is true not only for the institutions and practices of science but also for its intellectual content. Some scientists cling to the belief that one cannot possibly be a good scientist unless one is also a good citizen of the world, a liberal democrat, able to approach nature with heart and mind open. While neither Stark nor Lenard was perhaps as distinguished scientifically as their Nobel Prizes implied, we shouldn't make the mistake of imagining that something fundamental to their obnoxious political and social sympathies precluded them from continuing to function as scientists. The case is even more fraught for Pascual Jordan, one of the key figures in Bohr's circle as the Copenhagen interpretation of quantum mechanics was taking shape. Jordan concluded from the apparent demolition of objectivity by quantum theory that the 'liquidation' of the Enlightenment by the Third Reich was inevitable. What is more, according to the historian M. Norton Wise, Jordan's Nazi-leaning ideological views *informed* his physics, helping him to formulate aspects of quantum theory that were of genuine value and utility. 'It is necessary to make this point explicitly', says Wise,

> because we live with the tenacious myth that the acquisition of fundamental knowledge had to cease when scientists embraced Hitler. No real seekers after truth could also be pursuing Nazi political interests nor using those interests in the pursuit of knowledge. But of course they could, and did.

It is the other side of the same counterfeit coin to imagine that political interference in science happens only in dictatorships. Some is – or should be – unavoidable: science and technology need regulation, for example to ensure that certain ethical standards and responsibilities are met, and there is no obvious or consensual position on how far such constraints should extend: what to one professor is a reasonable demand might be repressive meddling to another. Funding, which can make or break a discipline, is highly politicized. Paul Forman's study of research in quantum electronic technologies in the post-war United States showed that, simply by how they choose to support particular research, governments 'can profoundly influence how scientists work – the questions they investigate, the methods they use, how they present their results'.

Democratically elected politicians have shown a readiness to challenge the autonomy, authority, integrity and validity of science. Not only do they sometimes find it expedient to ignore inconvenient advice from scientists – most egregiously, George W. Bush's resistance to the scientific consensus on climate change, although one might also adduce the persistent refusal of some Western governments to heed medical advice on drug-abuse policy – but they are not above rigging the evidence. Bush's Committee on Bioethics was chosen to engineer the advice on embryo research and stem-cell technology into a form that would play best to his constituency, while in 2007 the US House of Representatives Committee on Oversight and Government Reform concluded that 'the Bush Administration has engaged in a systematic effort to manipulate climate change science and mislead policymakers and the public about the dangers of global warming'. Turkey, a Muslim democracy, has recently brought its Academy of Sciences and its scientific funding agency under direct state control, a move that some say was prompted by a feeling in the government that the scientific community was too secular and liberal. Of course, one can argue in the mode of Churchill that democracy is the least bad of political systems for guarding against such meddling. That may well be true. But the cosy assumption that democracy guarantees good science and totalitarianism kills it finds little support in history. Moreover, if it is to engage with and sometimes oppose its political leaders, science needs the support of the rest of society. Scientists need legal protection from exclusion and persecution, a fact made all too evident in the politicization of climate science and biomedical research in the US, where some institutions have refused to defend individuals against litigation or intimidation from well-funded religious or climate-sceptic organizations.

Much has changed since Haberer delivered his rather damning judgement on the political and moral acumen of scientific communities – and that of Nazi Germany in particular – four decades ago, not least the growing awareness that science has a central role in tackling global crises such as environmental change and epidemic disease. But many scientists still cling to the shibboleth that their business is 'apolitical', a search for truth unsullied by worldly affairs. When the state does intrude on and interfere with science, scientists still struggle to find effective means of resistance. They can hardly carry the full blame for that; but history suggests that an aversion to political engagement will make such manipulation by the state all the easier. We should not wait for another dictatorship to cohere out of political and economic frustration and disenchantment before learning the lessons that the stories of Peter Debye, Max Planck and Werner Heisenberg can teach us.

Notes

Introduction

Nobel Prize-winner with dirty hands: Rispens (2006b).

willing helper of the regime: Rispens (2006a), translated in Altschuler (2006), 97.

Hitler's most important military: Rispens (2006b), translated in ibid., 96–7.

effective Aryan cleansing: Rispens (2006a), 180.

deprived of a hero: Eickhoff (2008), 146.

insufficiently resisted the limitations: press release from the University of Maastricht, 16 February 2006, 'Opgeroepen beeld moeilijk verenigbaar met voorbeeldfunctie UM', translated in Altschuler (2006), 98.

The Executive Board considers this picture: ibid.

recent evidence [was] not compatible: press release from Utrecht University, 16 February 2006, 'Universiteit Utrecht ziet af van naamgeving Debye'.

any action that dissociates Debye's name: press release from Cornell University, 2 June 2006.

persistent and virulent use: Walker (1995), 2.

One of the vital lessons: Gisevius (2009), 246.

Chapter 1

Prussia . . . could not afford: S. A. Goudsmit, 'The fate of German science', *Discovery*, August 1947, 239–43, here 242.

Like the majority of the professoriate: Beyerchen (1977), 1.

Respect for law: Heilbron (2000), 4.

by nature peaceful: ibid., 3.

The outside world: Pais (1991), 80.

What a glorious time: Heilbron (2000), 72.

the spotless purity: ibid., 5.

You can certainly be of: ibid., 85.

fortunate guess: Pais (1991), 85.

consists of a finite number of energy quanta: A. Einstein (1905), 'On a heuristic point of view concerning the production and transformation of light', *Annalen der Physik* **17**, 132–48. Transl. and reprinted in J. Stachel (ed.) (1998), *Einstein's Miraculous Year*, 178. Princeton University Press, Princeton.

as conservatively as possible: Heilbron (2000), 21.

private institutes and state institutes: Macrakis (1993), 34.

At the moment the outlook: Heilbron (2000), 88.

but there is one thing: Forman (1973), 163.

just like art and religion: ibid., 159.

For Germany the maintenance: ibid., 161–2.

the well-being of mankind: founding statement of John D. Rockefeller (1913): see http://centennial.rockefellerfoundation.org/values/top-twenty.

The nationalization of the great masses: A. Hitler (1926). *Mein Kampf*. Transl. in G. L. Mosse (ed.) (1966), 8.

Chapter 2

like a simple farm boy: Kumar (2008), 181.

One must probably introduce: Cassidy (2009), 115.

not only new assumptions: M. Born (1923), 'Quantentheorie und Störungsrechnung', *Naturwissenschaften* **11** (6 July), 537–42, here 542. Transl. in D. C. Cassidy (2007), 'Re-examining the crisis in quantum theory, Part 1: spectroscopy', paper presented at Max Planck Institute for the History of Science, 'Conference on the History of Quantum Physics', 2–6 July 2007, Berlin, 11.

The more precisely we determine: W. Heisenberg (1927), 'Über den anschaulichen Inhalt der quantentheoretischen. Kinematik und Mechanik', *Zeitschrift für Physik*, **43**, 172–98.

the meaninglessness of the causal law: Cassidy (2009), 178.

the concept – or the mere word: Forman (1971), 4.

denigrat[e] the capacity: ibid., 52.

solely from an inner need: ibid., 44.

liberation from the rooted prejudice: ibid., 88.

inherent irrationality: ibid., 107.

cannot escape the influence: ibid., 108.

The idea of such a crisis: ibid., 27.

the present crisis in mechanics: ibid., 62.

crisis in the foundations of mathematics: ibid., 60.

the present crisis in theoretical physics: ibid., 62.

Chapter 3

the loyal soldier and shield-bearer: Eickhoff (2008), 4.
that the best thing was for me: Debye (1965–6), unpaginated.
just a question of money: Debye (1962), unpaginated.
a charming boy who looked out: Davies (1970), 177.
We came to his house: Debye (1962).
Like the Viennese cafés: ibid.
the physicists [would] talk really about: Epstein (1965), 118.
even then an outstanding physicist: Davies (1970), 178.
Sommerfeld's most brilliant student: Kant (1997).
I feel myself to be very 'German': Eickhoff (2008), 2.
You should not even think: ibid., 16.
A good deal of German culture: Casimir (1983), 192.
made it difficult for me to identify: ibid.
whether, in the broad area: Davies (1970), 215.
Clever but lazy: Cassidy (2009), 190.
Debye had a certain tendency: ibid.
The beginning of something new: ibid., 103.

Chapter 4

Last week we received instructions: Hentschel (1996), 17.
in 1933 the barriers to state-sanctioned measures: Kershaw (2008), 40.
very moderately, tactfully: Hentschel (1996), 17.
only in retrospect is it so apparent: Beyerchen (1977), 69.
observers inside and outside: Rockefeller Archive RF RG 1.1 Projects, Series
    717, Folders 9–11, 'WET Diary' (W. E. Tisdale), 1 August 1934.
behind the pretty facade: Kurlander (2009), 23.
From 1929 on, it became more: Gisevius (2009), ix-x.
must bear a considerable measure: ibid., x.
overemphasis on individualism: ibid.
Even though liberal democrats: Kurlander (2009), 5.
much that is good: Cassidy (2009), 208.
although he had a loathing: R. Jungk (1956), letter to W. Heisenberg, 29
    December. Available at http://werner-heisenberg.physics.unh.edu/Jungk.
    htm.
Certainly opportunism and fear: Haberer (1969), 153.
religion, science and art: Kurlander (2009), 53.
On close examination: ibid., 47-8.

Weber's abrupt change of heart: ibid., 72–3.

The Third Reich was the product: ibid., 3.

One becomes ever more lonely: Mosse (ed.) (1966), 385.

The Jewish people: Hitler (1926). *Mein Kampf*. In ibid., 7.

pure moral religion stripped: Müller-Hill (1988), 90.

If you want to understand: Roth (2003), 210.

the aspects of Nazi ideology: Kurlander (2009), 18.

You simply do not conform: ibid., 164.

to prevent the worst excesses: ibid., 162.

lethal indifference: Kershaw (2008), 4.

Self-preservation is not a particularly: ibid., 148.

is the result of a very deep and general feeling: Rockefeller Foundation
    Archives, RF Officer Diaries, disk 16 (Warren Weaver), 24 May 1933, 82.

an influence for moderation: ibid.

They are almost frightened: ibid.

the new government has to give: ibid.

will not satisfy the crowd: ibid.

not particularly anti-Semitic: in N. Conrads (ed.) (2007), *Kein Recht, nirgends.
    Tagebuch vom Untergang des Breslauer Judentums 1933–1941*. Böhlau Verlag,
    Cologne.

one should acknowledge: ibid.

Civil servants who are not of Aryan descent: Hentschel (1996), 22.

The world-renowned intellectual freedom: Rockefeller Foundation Archive,
    RF Officer Diaries, disk 16 (Warren Weaver), 88.

I don't think he had time: ibid., 86.

temperamentally unfit: Heilbron (2000), 202.

all the troubles will be gone: Macrakis (1993), 53.

If 30 professors appealed: ibid., 68.

received the assurance that the government: Heilbron (2000), 154.

In the course of time: Sime (1996), 143.

anti-Jewish sneers and obscenities: Kramish (1986), 44.

do[es] not appear to fit in: N. Riehl & F. Seitz (1996). *Stalin's Captive*, 62.
    Chemical Heritage Foundation, Philadelphia.

I have serious doubts about [it]: ibid., 63.

I remember one distinguished member: P. Rosbaud, letter to L. Goudsmit,
    undated. In Samuel Goudsmit Papers, Series IV, Box 28, Folder 42.
    American Institute of Physics.

The general excuse was: ibid.

I noticed that the Germans: L. Szilard (1979), 'Leo Szilard: His Version of
    the Facts, Part II', *Bulletin of the Atomic Scientists* March, 56.

Many of them added: Rosbaud to Goudsmit, op. cit.

Would the populations: Kershaw (2008), 148.

After Hitler came to power: Heilbron (2000), 210.

all kinds of Jews: ibid.

That is not right: ibid.

it would be self-mutilation: ibid., 211.

uttered some commonplaces: ibid.

People say that I suffer: ibid.

and whipped himself: ibid.

said to Planck that he was not an anti-Semite: ibid., 213.

the consolidation of available forces: Macrakis (1993), 58.

utterly crushed: Heilbron (2000), 213.

disciplined thought must attend: ibid.

Nobody of the Nazi leaders: Rosbaud to Goudsmit, op. cit.

Chapter 5

I cannot allow it: Beyerchen (1977), 53.

My tradition requires of me: Heilbron (2000), 161.

stand before the pieces: Rockefeller Foundation Archive, RF Officer Diaries, disk 16 (Warren Weaver), 83.

a pathetic yet noble figure: ibid., 92.

What should I do?: Sime (1996), 142.

One was faced with the contradictory: Beyerchen (1977), 27.

I never did anything: F. Haber (1933), letter to C. Bosch, 28 December. Archives of the Max Planck Society, Haber Collection, Rep. 13, 911.

He was one of our own: Hentschel (1996), 78.

Haber remained true to us: Eickhoff (2008), 48.

old custom: Macrakis (1993), 70.

Haber has done a lot for science: ibid.

a device for justifying: Haberer (1969), 140.

benevolent protection: ibid., 147.

like a tree in the wind: Heim & Walker (2009), 3.

nothing is eaten as hot: Beyerchen (1977), 18.

equivalent to an act of sabotage: ibid., 19.

There is, of course, no chance: M. Born (1933), letter to A. Einstein, 2 June. In Born (2005), 115.

I would not have the nerve: ibid.

As regards my wife and children: ibid., 114.

My youngest son did not seem able: Beyerchen (1977), 26.

And how is mathematics in Göttingen: ibid., 36.

makes his defense of the situation: Rockefeller Foundation Archive, RF Officer Diaries, disk 16 (Warren Weaver), 85.

Only in the case of a few: ibid.

The watchword was that: Beyerchen (1977), 200.

Swastikas can be seen everywhere: Cassidy (2009), 222.

non-German spirit: ibid.

comrades under Hitler: ibid., 225.

With all respect for the freedom: Hentschel (1996), 60.

From now on, the question: Beyerchen (1977), 52.

As long as I have any choice: R. W. Clark (1973). *Einstein: The Life and Times*, 431. Hodder & Stoughton, London.

You know, I think, that I: Sime (1996), 140.

By your efforts, your racial: Walker (1995), 71.

Here they are making nearly the entirety: Cassidy (2009), 207.

I do not share your view: ibid., 207–8.

The Jews in Germany can thank refugees: Beyerchen (1977), 12.

I have learned that my unclear relationship: Hoffmann & Walker (2004), 52.

Although I am very thankful: ibid.

The Prussian Academy of Sciences heard: Einstein (1949), 82.

one of the most appalling experiences: Heilbron (2000), 158.

Therefore it is . . . deeply to be regretted: ibid., 159.

I am convinced that in the future: Haberer (1969), 114.

I hereby declare that I have never: Einstein (1954), 206.

I ask you to imagine yourself: Rowe & Schulman (2007), 274.

I am happy that you have nevertheless: ibid.

We had confidently expected: Einstein (1949), 85–6.

to counteract unjust suspicions: ibid., 84.

would have been equivalent: ibid., 86.

how you envisage your relations: ibid., 87.

The primary duty of an academy: ibid., 88.

When faced with a choice: Walker (1995), 92–3.

politically so worthless: Heilbron (2000), 159.

if Planck and Laue retain influence: ibid.

tactically a false step: ibid., 160.

Max Planck was one of the German professors: Haberer (1969), 128.

I think it was on the occasion: Beyerchen (1977), 1.

Chapter 6

the almost unanimous opinion: J. Z. Buchwald & A. Warwick (eds) (2004). *Histories of the Electron: The Birth of Microphysics*, 451. MIT Press, Cambridge, MA.

Jewish fraud: Beyerchen (1977), 93.

It was precisely the yearning: Mosse (ed.) (1966), 203.

seems already to indicate: Beyerchen (1977), 128.

arrogant delusion: Mosse (ed.) (1966), 205.

That influence has been even strengthened: ibid.

Jewry as such has outlived itself: R. Steiner (1971). *Gesammelte Aufsätze zur Literatur, 1884–1902*, 152. Rudolf Steiner Verlag, Basel.

scarcely believable filth: Heilbron (2000), 117.

a Jew of liberal international bent: Hentschel (1996), 1.

I admire Lenard as a master: ibid., 2.

los[ing] myself in such deep humorlessness: van Dongen (2007), 11.

enterprising business manager: Walker (1995), 6.

more and more of an anti-Semite: Rose (1998), 244.

appear to us as God's gifts: Hentschel (1996), 9.

is conditioned by the spiritual: Mosse (ed.) (1966), 206.

Respect for facts and aptitude: ibid., 205.

is focused upon its own ego: ibid., 207.

mix facts and imputations: ibid.

authentic creative work: ibid.

The scientist . . . does not exist: ibid., 206.

sets aside the concept of energy: ibid., 212.

something concerning the soul: ibid., 213.

His theory is not the keystone: ibid.

That which is called the crisis of science: Beyerchen (1977), 134.

To all of us minor figures: ibid., 66.

political battles call for different methods: Sime (1996), 157.

I hate them so much I must be close to them: ibid., 159.

was not a daredevil: ibid.

In Adolf Hitler we German: Walker (1995), 24–5.

From the beginning: Heilbron (2000), 170.

to make something sensible: ibid., 165.

playground for Catholics: Macrakis (1993), 98.

restricted circle . . . aristocratic splendour: ibid.

that the measures now being taken: Cassidy (2009), 229.

It is to the future that all of us: ibid., 231.

The world out there is really ugly: ibid.

spirit of Einstein's spirit: ibid., 247.
I now easily fall into a very strange state: ibid., 261.
bacterial carriers: ibid., 270.
still consists of Jews and foreigners: Hentschel (1996), 156.
Ossietzky of physics: Cassidy (2009), 270.
it was condemned by everyone: ibid., 271.
set the matter back in order: Beyerchen (1977), 159.
I do not approve of the attack: Cassidy (2009), 279.
I believe that Heisenberg: ibid., 280.
we may be able to get this man: Hentschel (1996), 176.
I would consider it proper: Beyerchen (1977), 163.
I never was sympathetic: Walker (2009), 358–9.
a political adviser and close friend: Peierls (1985), unpaginated.
I must mention him once: ibid.
after the war the names were: ibid.
America would have been discovered: Rose (1998), 270.
an unbreakable attachment: Cassidy (2009), 280.
remaining in Germany was apparently: ibid.
by seeing himself in such: ibid.
Heisenberg's notion of 'responsibility': Rose (1998), 260.
If you only understand theoretical physics: Beyerchen (1977), 166.
In the case of a purely scientific dispute: Hentschel (1996), 141.
Had he been less crazy: Heilbron (2000), 171.
their willingness and ability: Renneberg & Walker (1994), 10.
Prof. Prandtl is a typical scientist: ibid., 80.

Chapter 7

We visit Debye's institute: Rockefeller Foundation Archives RF RG 1.1 Projects, Series 717, Folders 9–11, memo from Warren Weaver, 21 January 1938.
The purpose of this institute: Kant (1996a), 228.
erected primarily for Einstein: Rockefeller Foundation Archives, op. cit., memo of 12 February 1930, 4.
land, buildings and equipment: ibid., memo of 16 April 1930.
prefer to stay in his own home: ibid., memo of 2 & 5 January 1931.
large gift: ibid., memo of 1 December 1936.
attempt to apply uniform: ibid.
The world of science is a world: *New York Times* 24 November 1936. In van Ginkel (2006), 14.

it is quite possible that the foundation: ibid.

adulterate[d] the spiritual coinage: Rockefeller Foundation Archives, op. cit., letter from F. Frankfurter to R. Fosdick, 24 November 1936.

the Nazi regime will not slacken: ibid., diary of Alan Gregg, 25 October 1933.

less ignorant and more moderate: ibid., diary of W. E. Tisdale, 16 June 1934.

the attitude of the present and future: ibid., memo from T. B. Appleget to M. Mason, 30 July 1934.

now given over entirely to work: ibid.

What might the physics institute: ibid.

to follow their example: Eickhoff (2008), 57.

within a year or two: Rockefeller Foundation Archives, op. cit., memo from W. E. Tisdale to W. Weaver, 1 August 1934.

Professor Debye is in my opinion: Kant (1996a), 235.

the best theoretician to be working: Eickhoff (2008), 46.

an arrangement of convenience: Rockefeller Foundation Archives, op. cit., diary of W. E. Tisdale, 12 June 1934.

Given on the understanding: Eickhoff (2008), 37.

Being a Dutch citizen at this time: van Ginkel (2006), 18.

something like a diplomatic status: Eickhoff (2008), 18.

I love Munich, and your presence: Kant (1996a), 236.

little man: Rockefeller Foundation Archives, RF Officer Diaries, disk 16 (Warren Weaver), 96.

he would take a page from the book: Rockefeller Foundation Archives, RF RG 1.1 Projects, Series 717, Folders 9–11, diary of W. E. Tisdale, 12 June 1934.

Debye seems to stand more firmly: ibid.

under the present regime: ibid., memo from W. E. Tisdale to W. Weaver, 1 August 1934.

the power to decide: ibid., diary of W. E. Tisdale, 29 August 1934.

the appeal leaves me quite cold: ibid., 4 September 1934.

negotiations are interminably slow: van Ginkel (1996), 10.

There is no doubt that under his leadership: Kant (1993).

an opportunity would be missed: ibid.

seriously damaged: Rockefeller Foundation Archives, RF RG 1.1 Projects, Series 717, Folders 9–11, letter to Mr Busser, Consul of the USA, Leipzig, 5 June 1935.

the only undepressed person: ibid., memo from W. E. Tisdale, 4 October 1935.

knows little, is pretty much worried: Eickhoff (2008), 60.

It soon became apparent: Kant (1996a), 240.

Debye was a very liberal director: ibid.

He said that the common notion: Rockefeller Foundation Archives, RF RG 1.1 Projects, Series 717, Folders 9–11, memo from Warren Weaver, 21 January 1938.

got the impression of a man: ibid.

I did this not for you: Macrakis (1993), 64.

it would be fair to say: Eickhoff (2008), 68.

I fear that these lines: S. L. Wolff (2011), 'Das Vorgehen von Debye bei dem Ausschluss der "jüdischen" Mitglieder aus der DPG'. In Hoffmann & Walker (eds) (2011), 118.

You obviously cannot swim: ibid.

I am disappointed that the society: Beyerchen (1977), 75.

have made only slight progress: Hentschel (1996), 178.

conspicuously lacked: ibid.

thoroughly informed about the position: ibid., 179.

gave an impossible incoherent talk: ibid.

it would be impossible for them: ibid.

to current requirements: ibid., 180.

would also have to provide: ibid.

Under the compelling prevailing circumstances: Hentschel (1996), 181.

turning point: Eickhoff (2008), 135.

In the mid-thirties all officials: van Ginkel (2006), 32.

The question was how would they sign it: Delbruck (1978).

will not have surprised you: van Ginkel (2006), preprint version provided by the author, 25.

I couldn't care less!: Eickhoff (2008), 90.

We must move away!: ibid., 23.

an immediate settlement of the Jewish question: S. L. Wolff, in Hoffmann & Walker (eds) (2011), 128.

quite considerable foreign currency receipts: Hentschel (1996), 181.

The DPG is still very backward: Hoffmann & Walker (2006a), unpaginated.

the handling of the Jewish question: ibid.

I can assure you: Wolff, op. cit., 132–3.

moral stamina: van Ginkel (2006), 33.

seemed to have a very high opinion: ibid.

by exercising [this] self-censorship: Eickhoff (2008), 96.

If you are thinking of appointing: van Ginkel (2006), 96

Debye engaged in an exclusive: Eickhoff (2008), 47.

I think that the racial issue: ibid., 96–7.

was not interested if [a] man: ibid., 24.

Today, among . . . contemporary Germans: Peierls (1985).

the Jewess endangers this institute: Sime (1996), 163.

He has, in essence, thrown me out: ibid., 185.

political objections exist: Hentschel (1996), 171.

When we last spoke: Sime (1996), 196.

The assistant we talked about: ibid., 202.

At the Dutch border: Rhodes (1986), 236.

You have made yourself as famous: Sime (1996), 205.

I very much hope that by now: Eickhoff (2008), 100.

connected with a survival mechanism: ibid., 148.

The last years have not passed: Reiding (2010), 291–2.

Debye moved, as a prominent scientist: ibid., 275.

Chapter 8

the field of atomic physics: Eickhoff (2008), 45.

I have seen my death!: G. Landwehr, Gottfried (1997). In A. Hasse (ed.), *Röntgen Centennial: X-rays in Natural and Life Sciences*, 7. World Scientific, Singapore.

because the question was entirely new: S. Quinn (1995). *Marie Curie: A Life*, 143. Simon & Schuster, New York.

Rutherford, this is transmutation: M. Howarth (1958). *Pioneer Research on the Atom*, 83–4. New World Publications, London.

some fool in a laboratory: Rhodes (1986), 44.

could be tapped and controlled: ibid.

It can even be thought: P. Curie (1905). 'Radioactive substances, especially radium'. 1903 Nobel Prize in Physics Lecture. Available at http://www.nobelprize.org/nobel_prizes/physics/laureates/1903/pierre-curie-lecture.pdf.

always tired without being exactly ill: Quinn, op. cit., or Reid (1974), *Marie Curie*.

in a very inflamed: Rhodes (1986), 45.

It was quite the most incredible: ibid., 49.

to get some experience: ibid., 52.

To change the hydrogen: ibid., 140.

We can only hope that: ibid., 141.

there was something quite new: ibid., 162.

anyone who looked for a source: ibid., 27.

time cracked open: ibid., 13.

inner rectitude: Sime (1996), 37.

possess[es] a special gift: ibid., 25.

Amazons are abnormal: ibid., 26.

Oh, I thought you were a man!: ibid., 33.

I value my personal freedom: ibid., 157.

Perhaps you can suggest: Rhodes (1986), 253.

pretentious . . . exactly of [the] sort: P. Rosbaud, letter to S. Goudsmit. In Samuel Goudsmit Papers, Series IV, Alsos Mission: Box 28, Folder 42, 15. American Institute of Physics.

We permit ourselves to direct your attention: Kramish (1986), 53.

Chapter 9

for military technological ends: Eickhoff (2008), 102.

In response to Dr Telschow's: ibid.

There was no question of surrender: ibid., 105.

stay at home and write a book: Davies (1970), 209.

knowledge and approaches for military objectives: Eickhoff (2008), 103.

Until now the institute: Rockefeller Foundation Archives, RF RG 1.1, Series 200D, Box 136, Folders 1677–8, letter from P. Debye to W. E. Tisdale, 7 October 1939.

Dr Peter J. W. Debye: van Ginkel (2006), 41.

In my view there is also the possibility: Eickhoff, 106.

for the duration of the war: ibid.

I had to give up all these beautiful laboratories: Debye (1964), 34.

German physics has lost: Hoffmann (2005), 311.

the large-scale research: ibid., 317.

increasing acceptance of physics: ibid., 319.

our technical development: R. W. Gerard (1949), 'The scientific reserve', *Bulletin of the Atomic Scientists* October, 276–80, here 277.

3,000 more physicists could perhaps: Hoffmann (2005), 319.

If someone wanted to research: Beyerchen (1977), 191.

Some of my colleagues think: van Ginkel (2006), 53.

suspicious of him: ibid.

extremely mercenary: Eickhoff (2008), 129.

not loyal even to the field: van Ginkel (2006), 102.

could not be trusted: ibid.

has no emotional reaction: ibid., 101.

Einstein advised that he has never: ibid., 55.

Einstein said that he does not believe: ibid.

unspoiled soul: van Ginkel (2006), 57.

he found Einstein in fact to be: interview with Caroline Debye, 1 April

1970. Debye archives of the Regionaal Historisch Centrum Limburg, Maastricht, 8.

secret purpose: van Ginkel (2006), 54.

since I had no possibility: A. Einstein, letter to J. G. Kirkwood, 17 June 1940. In Albert Einstein Archives, Hebrew University of Jerusalem, Israel, 9–150.

Those suspicions are entirely groundless: P. Debye, letter to A. Einstein, 12 June 1940. In Albert Einstein Archives, Hebrew University of Jerusalem, Israel, 9–145.

symptomatic of the kind of hysteria: Rockefeller Foundation Archives, op. cit., memo from W. Weaver, 13 June 1940.

has every confidence that Debye: ibid.

the result of Jewish prejudice: van Ginkel (2006), 58.

declined to attend: Rockefeller Foundation Archives, op. cit., memo from W. Weaver.

some of Debye's colleagues: ibid.

Debye is a man of high character: van Ginkel (2006), 59.

The army has made this move: Rockefeller Foundation Archives, RF Officer Diaries, disk 16 (Warren Weaver), memo of 6 February 1940, 19–20.

the chances are infinitely against: Eickhoff (2008), 124.

I could have done so: ibid., 140.

should not be entrusted: van Ginkel (2006), 127.

took it with immense good humour: van Ginkel (2006), 69.

Accepting the Baker lectureship: Eickhoff (2008), 111, n.6.

Debye has now practically: ibid., 114.

keep the back door open: ibid., 154.

My letters remain unanswered: ibid., 117.

was caught between several: ibid., 131.

It is curious that about four weeks ago: ibid., 118.

against the wishes: ibid.

as soon as you are again able: van Ginkel (2006), 73.

Professor Debye states that he is: Eickhoff (2008), 119.

Peter Debye continued the negotiations: van Ginkel (2006), 71.

strange: S. L. Wolff, private communication.

Hilde and Maida prefer: Eickhoff (2008), 25.

so he won't lose everything: ibid.

that his wife was born in Germany: ibid., 120.

royal salary: Rispens (2006b).

It would be undesirable: Eickhoff (2008), 122.

somewhere in Czechoslovakia: van Ginkel (2006), 82.

it was common for scientists: Hoffmann & Walker (2006a).

Many nasty things have been said: Delbrück (1978).
Going away without any kind of security: ibid.

Chapter 10

more than ten times as powerful: Heim, Sachse & Walker (eds) (2009), 343.
in the foreseeable future: Cassidy (2009), 321.
enormous significance: ibid.
success can be expected shortly: ibid.
once in operation: Hentschel (1996), 300.
one can say that the first time: Bernstein (ed.) (2007), 121.
it was from September: Irving (1967), 114.
hitherto unknown explosive: Walker (1995), 158.
I thank you for the rehabilitation: Heim, Sachse & Walker (eds) (2009), 359.
probably have become critical: C. F. von Weizsäcker (1991). 'Die bombe war
    zu teuer', *Die Zeit* 24 May, 18. Available at http://www.zeit.de/1991/17/
    die-bombe-war-zu-teuer.
the sun will continue to shine: Cassidy (2009), 359.
Germany needs me: ibid., 370.
I am in a rather sad and violent: S. Goudsmit (1948), letter to P. Rosbaud,
    April. In Samuel Goudsmit Papers, Series IV, Alsos Mission: Box 28, Folder
    43, 21. American Institute of Physics.
Don't think that Heisenberg: P. Rosbaud (1948), letter to S. Goudsmit, April,
    ibid.
I doubt that I or most of the physicists: Cassidy (2009), 355.

Chapter 11

That's impossible!: Bernstein (ed.) (2001), 53.
I don't think they know the real Gestapo: ibid., 78.
Things can't go on like this: ibid.
It won't do: ibid.
These people have detained us firstly: ibid., 81.
The first atomic bomb has been dropped: see http://news.bbc.co.uk/
    onthisday/hi/dates/stories/august/6/newsid_3602000/3602189.stm.
You're just second-raters: Bernstein (ed.) (2001), 116.
All I can suggest is that: ibid.
then we are in luck: Frank (ed.) (1993), 93.
I think it is dreadful: Bernstein (ed.) (2001), 117.

I believe the reason we didn't do it: ibid., 122.
Our entire uranium research: Beyerchen (1977), 197.
After that day we talked much: Bernstein (ed.) (2001), 352–3.
I don't intend to make any war physics: P. Rosbaud (1945), letter to S. Goudsmit, 5 August, 5. In Samuel Goudsmit Papers, Series IV, Alsos Mission: Box 28, Folder 42. American Institute of Physics.
Germany must not lose the war: P. Rosbaud (1945), letter to S. Goudsmit, undated fragment. In ibid.
He could not and probably: ibid.
I went to my downfall: Bernstein (ed.) (2001), 134.
History will record: ibid., 138.
It seems paradoxical that: Jungk (1958), 105.
what these scientists felt they needed most: Walker (1995), 241.
betrayed: Dörries (ed.) (2005), 52.
I want to thank you very much: W. Heisenberg (1956), letter to R. Jungk, 17 November. Available at http://werner-heisenberg.physics.unh.edu/Jungk.htm.
Heisenberg did not simply withhold: Dörries (ed.) (2005), 53.
Dr Hahn, Dr von Laue and I: R. N. Anshen (1986). *Biography of an Idea*, 71. Moyer Bell, Mount Kisco, NY.
tragically absurd: Walker (1995), 267.
to keep order in those small corners: Heisenberg (1971), 167.
uncontaminated science: J. Medawar & Pyke (2001). *Hitler's Gift: The True Story of the Scientists Expelled by the Nazi Regime*, 171. Arcade, New York.
courageous polyphony: Dörries (ed.) (2005), 37.
the true goal of the visit: Walker (1995), 257.
Obviously Professor Bohr: Karlsch & Walker (2005), 17.
Heisenberg and Weizsäcker sought: Niels Bohr Archive, documents released 6 February 2002, Document 6 (drafted by Margrethe Bohr). Available at www.nba.nbi.dk/papers/docs/cover.html.
biological necessity: Walker (1995), 149.
appearance in Denmark in 1941: Hentschel (1996), 334.
I have heard very peculiar things: Dörries (ed.) (2005), 39.
It is amazing, given that the Danes: W. Heisenberg (1941), letter to his wife Elisabeth, September. Available at http://werner-heisenberg.physics.unh.edu/copenhagen.htm.
Because I knew that Bohr: W. Heisenberg (1956), letter to R. Jungk, op. cit.
Everything I am writing here: ibid.
Personally, I remember every word: N. Bohr, letter to W. Heisenberg, undated.

Niels Bohr Archive, documents released 6 February 2002, Document 1. Available at www.nba.nbi.dk/papers/docs/d01tra.htm.html.

you informed me that it was: ibid., Document 7, www.nba.nbi.dk/papers/docs/d07tra.htm.html.

deeply mistaken: Dörries (ed.) (2005), 36.

They didn't need my help: Rhodes (1986), 525.

He made the enterprise: ibid., 524.

History legitimizes Germany: G. Kuiper (1945), report to Major Fischer, 30 June. In University of Arizona Library, Kuiper Papers, Box 28.

From the very beginning: Heisenberg (1947), 214.

The actual givens of the situation: W. Heisenberg (1956), letter to R. Jungk, op. cit.

Obviously we were not *fully* aware: Hentschel (1996), lxxxvi.

spared the decision: Heisenberg (1947), 214.

The official slogan of the government: Cassidy (2009), 305.

fought the Nazis not because: Goudsmit (1947), 115.

presumably around 10–100 kilograms: Walker (2009), 353.

I thought that one needed only: Bernstein (ed.) (2001), 117.

This statement shows that at this point: ibid.

But tell me why you used to tell me: ibid., 118.

about the size of a pineapple: W. Heisenberg, letter to S. Goudsmit, 3 October 1948, 3. In Samuel Goudsmit Papers, Box 10, Folder 95. American Institute of Physics.

This statement of course caused: ibid.

makes it crystal clear: Karlsch & Walker (2005), 17.

as far as they got: M. Walker, personal correspondence.

We could feel satisfied with the hope: Heisenberg (1947), 214–15.

dreamy wish: Walker (2009), 346.

We always thought we would: Bernstein (ed.) (2001), 117.

he stands for professional excellence: Heilbron (2000), 216.

What he did during the Nazi period: ibid.

physicists have a right to know: Hoffmann (2005), 294.

that they had done everything possible: ibid., 324.

the inborn conceit: Frank (ed.) (1993), 168.

never reckoned with the possibility: Renneberg & Walker (eds) (1994), 252.

Most of our old friends: Rose (1998), 309.

we had enough trouble: Hentschel (2012), 332.

It was one of the most depressing: ibid., 318.

absolutely bitter, negative: ibid., 322.

the tough momentary situation: ibid., 355.

It is a difficult problem: ibid., 356.

it seemed that the only thing: ibid., 322.
He still goes on defending: Rose (1998), 311.
I would like to apologize: Heim, Sachse & Walker (2009), 7.
The conditioning of German culture: Rose (1998), 75.
it remains incomprehensible: Dörries (ed.) (2005), 41.
With your completely different: ibid., 43.
It is, in most cases, morally wrong: S. Goudsmit (1947), 'German scientists in army employment I: the case analysed', *Bulletin of the Atomic Scientists* February, 64.
The documents cited in *Alsos*: Morrison (1947), 365.
brave and good men like Laue: ibid.
Ever since war between: Laue (1948).
honest, solid scientific investigation: ibid.
what unutterable pain: ibid.
How careful one must be: ibid.
We recommend as the foundation: ibid.
Many of the most able: Morrison (1948).
but many a famous German: ibid.
is not helping Germany: Hentschel (1996), 402.
Someone should force a man: ibid., 334.
It was clear to me: Hoffmann (2005), 325.
Today I know that it was not only stupid: Hentschel (1996), 334.
We all knew that injustice: Sime (1996), 364.
for trying to make us understand: ibid.
a physicist who never lost her humanity: ibid., 380.

Chapter 12

You alone kept me from concluding: Vonnegut (1961), 75.
we are what we pretend to be: ibid., v.
It wasn't that Helga and I: ibid., 28–9.
a man who served evil too: ibid., xii.
Under totalitarianism: Gisevius (2009), 44–5.
I have done that: Rose (1998), 7.
But you see, it's all terribly simple: Mansel (1970), 216.
To the end his generosity: Williams (1975), 47.
In his own eyes he had got through: Eickhoff (2008), 138.
not a lovable character: Epstein (1965), 79.
not a man of the very highest: ibid.
I saw through him: ibid., 86.

Debye was also a profoundly: ibid.

refused to be browbeaten: Eickhoff (2008), 144.

great personal courage: ibid., 145.

striking lack of political interest: ibid., 141.

I never found in Debye: van Ginkel (2006), 90.

is not based on sound historical: Schultz (2006).

If I had realized the consequences: van Ginkel (2006), 148.

We believe you have done: the Debye family (2006), letter to the University
    Board of Utrecht University, 20 June.

On World War II we Dutch: D. Hartmann & J. van Turnhout (2006), 'Zestig
    Jaar later: niemand is veilig', letter in *Het Parool*, 29 August. See http://
    home.kpn.nl/i.geuskens/peterdebye/DebyeTurn.htm.

an ordinary man: Hoffmann & Walker (2006a).

Based on the information to date: Cornell University press release, 2 June
    2006. In van Ginkel (2006), 149.

Debye took on positions of administration: R. Hoffmann (2006), *Chemical
    & Engineering News* 24 July, 6.

Not to despair and always be ready: van Ginkel (2006), 3.

opportunistic behaviour: Eickhoff (2008), 154.

have the courage to relinquish: Dörries (ed.) (2005), 34.

After the war, Debye made no apology: R. Hoffmann (2006), op. cit.

money-business: Cassidy (2009), 64.

I knew . . . if we Germans did not succeed: W. Heisenberg, letter to S.
    Goudsmit, 5 January 1948, 4. In Samuel Goudsmit Papers, Box 10, Folder
    95. American Institute of Physics.

For us there remains nothing: Rose (1998), 286.

Heisenberg knew he was working: Walker (1995), 172.

not one of the German colleagues: S. Goudsmit, letter to W. Heisenberg,
    20 September 1948, 2. In Samuel Goudsmit Papers, Box 10, Folder 95.
    American Institute of Physics.

the truth was not that the scientists: Beyerchen (1977), 207.

It is always my custom to ask: van Ginkel (2006), 49.

Debye comments that Hitler accomplished: Eickhoff (2008), 109.

we have far-reaching pioneering work: F. Dürrenmatt (1964). *The Physicists*,
    transl. J. Kirkup, 54.

new and inconceivable forces: ibid. 53. Jonathan Cape, London.

Let us not forget that totalitarianism: Gisevius (2009), 42.

I felt I had been a coward: Casimir (1983), 191–2.

was not opposition at all: Beyerchen (1977), 206–7.

the motive is not to honor: Reiding et al. (2008), unpaginated.

Epilogue

atheist extremism: address of Benedict XVI, Palace of Holyroodhouse, Edinburgh, 16 September 2010. Available at http://www.vatican.va/holy_father/benedict_xvi/speeches/2010/september/documents/hf_ben-xvi_spe_20100916_incontro-autorita_en.html.

accustomed to regarding matters: Hentschel (2012), 334.

an idealization of science: Haberer (1969), 2.

The real issue involves how: ibid., 152–3.

Whether they support the regime: Walker (ed.) (2003), 59–60.

people who have to deal with black employees: see http://news.bbc.co.uk/1/hi/7052416.stm.

the hounding, by what can only be described: R. McKie (2007), 'Disgrace: how a giant of science was brought low', *Observer* 21 October. See http://www.guardian.co.uk/uk/2007/oct/21/race.research.

scientific importance: Hoffmann & Walker (eds) (2012), 390.

The challenge to the understanding: Douglas (2003), 60.

is infused with problems: Haberer (1969), 299.

scientific leadership has tended: ibid., 303.

one of the blackest comedies: Rhodes (1986), 529.

He scolded us like schoolboys: ibid., 529–30.

We regard it as proper: J. R. Oppenheimer (1989). *Atom and Void: Essays on Science and Community*, 74–5. Princeton University Press, Princeton.

In most scientific study: Haberer (1969), 258.

We hear different language: ibid., 252.

elude specific meaning: ibid., 261.

It must be possible both to respect: Walker (1995), 271.

Many of the ways: Macrakis (1993), 4.

The failure of scientists: Haberer (1969), 311.

disgraceful role played by a few scientists: J. Rotblat (1995), Nobel Peace Prize 1995 Lecture. Available at http://www.pugwash.org/award/Rotblatnobel.htm

When it comes to nuclear weapons: ibid.

You are doing fundamental work: ibid.

Looking back now, this unique conference: P. Berg (2004), 'Asilomar and recombinant DNA', article available at http://www.nobelprize.org/nobel_prizes/chemistry/laureates/1980/berg-article.html.

unfettered pursuit of this research: ibid.

First and foremost, we gained: ibid.

The history of science in totalitarian societies: Walker (ed.) (2003), 32.

no single ideology: ibid., 58.

the Kaiser Wilhelm Society was an integral part: Heim, Sachse & Walker (eds) (2009), 4.

they will adapt to political and social changes: Renneberg & Walker (eds) (1994), 310.

I cannot find any reason: ibid., 311.

No political regime has ever tried: Walker (ed.) (2003), 58.

Stalin left his nuclear physicists alone: T. Judt (2010). *Postwar: A History of Europe Since 1945*, 174. Vintage, London.

It is necessary to make this point: Renneberg & Walker (eds) (1994), 244.

can profoundly influence how scientists work: Heim, Sachse & Walker (eds) (2009), 13.

the Bush Administration has engaged: US House of Representatives Committee on Oversight and Government Reform (2007), 'Political interference with climate change science under the Bush Administration', December. Executive Summary, i.

Bibliography

H. D. Abruña (2006). 'Peter Debye', *Chemical and Engineering News* July 24, 4–6.

H. Albrecht (1993). 'Max Planck: Mein Besuch bei Adolf Hitler' – Anmerkungen zum Wert einer historischen Quelle, in H. Albrecht (ed.), *Naturwissenschaft und Technik in der Geschichte* 41–63. Verlag für Geschlichte der Naturwissenschaft und Technik, Stuttgart.

G. C. Altschuler (2006). 'The convictions of Peter Debye', *Daedalus*, Fall, 96–103.

J. Baggott (2009). *Atomic*. Icon, London.

J. Bernstein (ed.) (2001). *Hitler's Uranium Club: The Secret Recordings at Farm Hall*, 2nd edn. Copernicus, New York.

A. D. Beyerchen (1977). *Scientists under Hitler: Politics and the Physics Community in the Third Reich*. Yale University Press, New Haven.

M. Born (2005). *The Born–Einstein Letters*. Macmillan, London.

P. Bridgman (1948). 'Scientists and social responsibility', *Bulletin of the Atomic Scientists* **4(3)**, 69–72.

H. Casimir (1983). *Haphazard Reality*. New York.

D. C. Cassidy (2009). *Beyond Uncertainty: Heisenberg, Quantum Physics, and the Bomb*. Bellevue Literary Press, New York.

J. Cornwell (2003). *Hitler's Scientists*. Viking, London.

M. Davies (1970). 'Peter Joseph Wilhelm Debye: 1884–1966', *Biographical Memoirs of Fellows of the Royal Society* **16**, 175–232.

P. Debye (1965–66). Oral interview conducted by D. M. Kerr and L. P. Williams, 22 December 1965, 20 January 1966, 6 June 1966 & 16 June 1966. Kroch Library, Cornell University, Archive 13–6–2282 trsc.5216–5218.

P. Debye (1962). Oral interview conducted by T. Kuhn & G. Uhlenbeck, 3 May 1962. Available at www.aip.org/history/ohilist/4568_1.html.

P. Debye (1964). Oral interview conducted by H. Zuckerman. Oral History Research Office, Butler Library, Columbia University, New York, Box 20, Room 801.

M. Delbrück (1978). Oral interview conducted by C. Harding, 14 July–11

September. Archives of the California Institute of Technology, Pasadena. Available at oralhistories.library.caltech.edu/16/1/OH_Delbruck_M.pdf.

J. van Dongen (2007). 'Reactionaries and Einstein's Fame: "German Scientists for the Preservation of Pure Science," Relativity, and the Bad Nauheim Meeting'. *Physics in Perspective* **9**, 212–30.

M. Dörries (ed.) (2005). *Michael Frayn's Copenhagen in Debate.* University of California Press, Berkeley.

H. Douglas (2003). 'The moral responsibilities of scientists'. *Amercian Philosophical Quarterly* **40**, January, 59–68.

M. Eickhoff (2008). *In the Name of Science? P. J. W. Debye and His Career in Nazi Germany*, transl. P. Mason. Aksant, Amsterdam.

A. Einstein (1954). *Ideas and Opinions.* Bonanza Books, New York.

A. Einstein (1949). *The World as I See It.* Philosophical Library, New York.

P. Epstein (1965). Oral interview conducted by Alice Epstein, Pasadena, California, beginning 22 November 1965. Oral History Project, California Institute of Technology. Available at http://oralhistories.library.caltech.edu/73/.

P. Forman (1971). 'Weimar culture, causality, and quantum theory, 1918–1927: adaptation by German physicists and mathematicians to a hostile intellectual environment', *Historical Studies in the Physical Sciences*, ed. R. McCormmach, 1–115. University of Pennsylvania Press, Philadelphia.

P. Forman (1973). 'Scientific internationalism and the Weimar physicists: the ideology and its manipulation in Germany after World War I', *Isis* **64**, 151–80.

C. Frank (ed.) (1993). *Operation Epsilon: The Farm Hall Transcripts.* Institute of Physics Publishing, Bristol.

H. B. Gisevius (2009). *'Valkyrie': An Insider's Account of the Plot to Kill Hitler.* Da Capo, Philadelphia. Originally published as *To the Bitter End*, Da Capo, Cambridge, Ma., 1947.

S. A. Goudsmit (1947). *Alsos.* Henry Schuman, New York.

S. A. Goudsmit (1921–79). Samuel A. Goudsmit Papers. American Institute of Physics. Available at http://www.aip.org/history/nbl/collections/goudsmit/.

J. Haberer (1969). *Politics and the Community of Science.* Van Nostrand Reinhold Co., New York.

J. L. Heilbron (2000). *The Dilemmas of an Upright Man: Max Planck and the Fortunes of German Science*, 2nd edn. Harvard University Press, Cambridge, Ma.

S. Heim, C. Sachse & M. Walker (eds) (2009). *The Kaiser Wilhelm Society under National Socialism.* Cambridge University Press, Cambridge, 2010.

W. Heisenberg (1947). 'Research in Germany on the technical application of atomic energy', *Nature* **160**, 211–215.

W. Heisenberg (1971). *Physics and Beyond: Encounters and Conversations*, transl. A. J. Pomerans. George Allen & Unwin, London.

K. Hentschel (ed.) (1996). *Physics and National Socialism: An Anthology of Primary Sources*, transl. A. M. Hentschel. Birkhauser Verlag, Basel.

K. Hentschel (2012), 'Distrust, bitterness, and sentimentality: On the mentality of German physicists in the immediate post-war period'. In Hoffmann & Walker (2012).

R. Hoffmann (2006), 'Peter Debye', *Chemical and Engineering News* July 24, Volume 84, Number 30, pp. 4–6.

D. Hoffmann (1997). 'Max Planck (1858–1947): Leben– Werk–Persönlichkeit', in *Max Planck: Vorträge und Ausstellung zum 50. Todestag*. Max-Planck-Gesellschaft, Munich.

D. Hoffmann (2005). 'Between autonomy and accommodation: the German Physical Society during the Third Reich', *Physics in Perspective* 7(3), 293–329.

D. Hoffmann & M. Walker (eds) (2011). *'Fremde' Wissenschaftler im Dritten Reich. Die Debye-Affäre im Kontext*. Wallstein Verlag, Göttingen.

D. Hoffmann & M. Walker (2004). 'The German Physical Society under National Socialism', *Physics Today* 57(12), 52–8.

D. Hoffmann & M. Walker (2006a). 'Peter Debye: a typical scientist in an untypical time'. Available at www.dpg-physik.de/dpg/gliederung/fv/gp/debye_en.html.

D. Hoffmann & M. Walker (eds) (2006b). *Physiker zwischen Autonomie und Anpassung*. Wiley, Berlin.

D. Hoffmann & M. Walker (eds) (2012). *The German Physical Society in the Third Reich*. Cambridge University Press, Cambridge.

J. Hughes (2009). 'Making isotopes matter: Francis Aston and the mass-spectrograph', *Dynamis* 29, available at http://dx.doi.org/10.4321/S0211-95362009000100007.

S. E. Hustinx & C. Bremen (eds) (2000). *Pie Debije – Peter Debye 1884–1966*. Gardez!, St Augustin.

D. Irving (1967). *The Virus House*. Kimber, New York.

L. E. Jones (1988). *German Liberalism and the Dissolution of the Weimar Party System 1918–1933*. University of North Carolina Press, Chapel Hill.

R. Jungk (1958). *Brighter than a Thousand Suns*. Harcourt, Brace & Co., New York.

H. Kant (1997). 'Peter Debye (1884–1966)', in *Die grossen Physiker* Vol. 2, p. 263–75. Beck, Munich.

H. Kant (1996a). 'Albert Einstein, Max von Laue, Peter Debye und das Kaiser-Wilhelm-Institut für Physik in Berlin (1917–1939)', in B. vom Brocke & H. Laitko (eds), *Die Kaiser-Wilhelm-/Max-Planck-Gesellschaft und ihre Institute*, 227–43. Walter de Gruyter, Berlin.

H. Kant (1996b). 'Peter Debye und die Deutsche Physikalische Gesellschaft'. In D. Hoffmann, F. Bevilacqua & R. Stuewer (eds), *The Emergence of Modern Physics*, Proceedings of a Conference Commemorating a Century of Physics, Berlin, 22–24 March 1995, 505–20. Università degli Studi di Pavia.

Horst Kant (1993). 'Peter Debye und das Kaiser-Wilhelm-Institut für Physik in Berlin'. In H. Albrecht (ed.), *Naturwissenschaft* und *Technik in der Geschichte: 25 Jahre Lehrstuhl für Geschichte der Naturwissenschaft und Technik am Historischen Institut der Universität Stuttgart*. Verlag für Geschichte der Naturwissenschaften und der Technik, Stuttgart.

R. Karlsch & M. Walker (2005). 'New light on Hitler's bomb', *Physics World* June, 15–18.

I. Kershaw (2008). *Hitler, the Germans, and the Final Solution*. Yale University Press, New Haven.

A. Kramish (1986). *The Griffin*. Macmillan, London.

T. S. Kuhn & G. Uhlenbeck (1962). Interview with Dr Peter Debye, 3 May. AIP oral history http://www.aip.org/history/ohilist/4568_1.html.

M. Kumar (2008). *Quantum: Einstein, Bohr and the Great Debate about the Nature of Reality*. Icon, Cambridge.

E. Kurlander (2009). *Living with Hitler*. Yale University Press, New Haven.

M. von Laue (1948). 'The wartime activities of German scientists', *Bulletin of the Atomic Scientists* April, 103.

A. Long (1967). 'Peter Debye: An Appreciation', *Science* **155**, 979.

K. Macrakis (1993). *Surviving the Swastika: Scientific Research in Nazi Germany*. Oxford University Press, New York.

G. L. Mosse (ed.) (1966). *Nazi Culture: Intellectual, Cultural and Social Life in the Third Reich*. Grosset & Dunlap, New York.

P. Morrison (1947). 'Alsos: The story of German science', *Bulletin of the Atomic Scientists* December, 354–65.

P. Morrison (1948). 'A reply to Dr von Laue', *Bulletin of the Atomic Scientists* April, 104.

B. Müller-Hill (1988). *Murderous Science*. Oxford University Press, Oxford.

Niels Bohr Archive, http://www.nba.nbi.dk/

J. R. Oppenheimer (1989). *Atom and Void: Essays on Science and Community*. Princeton University Press, Princeton.

A. Pais (1991). *Niels Bohr's Times*. Clarendon, Oxford.

R. Peierls (1985). Interview with Mark Walker, 7 April. Niels Bohr Library & Archives, American Institute of Physics, College Park, Maryland. Available at http://www.aip.org/history/ohilist/4819.html.

T. Powers (1993). *Heisenberg's War: The Secret History of the German Bomb*. Da Capo, Cambridge, Ma.

G. Rammer (2012), '"Cleanliness among our circle of colleagues": the

German Physical Society's policy towards its past', in Hoffmann & M. Walker (eds) (2012).

J. Reiding (2010). 'Peter Debye: Nazi collaborator or secret opponent?', *Ambix* 57, 275–300.

J. Reiding, E. Homberg, K. van Berkel, L. Dorsman & M. Eickhoff (2008). 'Discussiedossier over Debye', *Studium* 4, 269–86. Gewina (Dutch Society for the History of Science and Universities).

M. Renneburg & M. Walker (eds) (1994). *Science, Technology and National Socialism*. Cambridge University Press, Cambridge.

R. Rhodes (1986). *The Making of the Atom Bomb*. Simon & Schuster, New York.

S. I. Rispens (2006a). *Einstein in Nederland: Een Intellectuelle Biographie*. Ambo, Amsterdam.

S. Rispens (2006b). 'Peter Debye, nobelprijswinnaar met vuile handen', *Vrij Nederland* 21 January 2006. Available at http://www.vn.nl/Archief/Wetenschapmilieu/Artikel-Wetenschapmilieu/Peter-Debye-nobelprijswinnaar-met-vuile-handen.htm.

Rockefeller Foundation Archive, Tarrytown, New York.

P. L. Rose (1998). *Heisenberg and the Nazi Atomic Bomb Project: A Study in German Culture*. University of California Press, Berkeley.

J. Roth (2003). *What I Saw: Reports from Berlin, 1920–1933*, ed. & transl. M. Hofmann. Granta, London.

D. E. Rowe & R. J. Schulmann (2007). *Einstein on Politics: His Private Thoughts and Public Stands on Nationalism, Zionism, War, Peace, and the Bomb*. Princeton University Press, Princeton.

C. Sachse & M. Walker (eds) (2005). *Politics and Science in Wartime: Comparative International Perspectives on the Kaiser Wilhelm Institutes*. University of Chicago Press, Chicago.

W. Schulz (2006). 'Nobel laureate is accused of Nazi collaboration', *Chemical and Engineering News* 6 March, 19.

R. L. Sime (1996). *Lise Meitner: A Life in Physics*. University of California Press, Berkeley.

J. Stark (1938). 'The pragmatic and the dogmatic spirit in physics', *Nature* 141, 770–72.

M. Szöllösi-Janze (ed.) (2001). *Science in the Third Reich*. Berg, Oxford.

G. van Ginkel (2006). *Prof. Peter J. W. Debye (1884–1966) in 1935–1945: An Investigation of Historical Sources*. RIPCN, the Netherlands.

G. van Ginkel & C. Bremen (2007). 'Die Kontroverse um "Aachens berühmtesten Schüler": Peter Debye'. *Zeitschrift des Aachener Geschichtsvereins* Band 109, 101–150.

M. Walker (1989). *German National Socialism and the Quest for Nuclear Power 1939–1949*. Cambridge University Press, Cambridge.

M. Walker (1990). 'Heisenberg, Goudsmit and the German atomic bomb', *Physics Today* **43(1)**, 52–60.

M. Walker (1995). *Nazi Science: Myth, Truth and the German Atomic Bomb.* Plenum, New York.

M. Walker (2009). 'Nuclear weapons and reactor research at the Kaiser Wilhelm Institute for Physics', in Walker & Hoffmann (eds) (2012).

M. Walker (ed.) (2003). *Science and Ideology: A Comparative History.* Routledge, London.

Werner Heisenberg pages, http://werner-heisenberg.physics.unh.edu/.

J. W. Williams (1975). 'Peter Joseph Wilhelm Debye', *Biographical Memoirs of the National Academy of Sciences.* National Academy of Sciences, Washington, DC.

Image Credits

Index